Hawks & Owls
OF EASTERN NORTH AMERICA

Chris G. Earley

Dedication

TO NATHAN, FOR ALL OF OUR PAST OWL PROWLS.
AND TO SKYE, for all our future ones.

Credits

PHOTOGRAPHERS: TONY BECK, KARL EGRESSY, SCOTT FAIRBAIRN, JIM
Flynn, Brandon Holden, Wayne Lynch, Robert McCaw, John Reaume,
Brain E. Small, Brian K. Wheeler

THE RANGE MAPS THROUGHOUT THIS BOOK WERE GENEROUSLY PROVIDED
by WILDSPACE™ 2002. WILDSPACE™: digital hemispheric range maps for
the breeding birds of Canada. Canadian Wildlife Service, Ontario Region,
Ottawa Ontario, Canada.

SPECIAL THANKS to Karl Konze and Jim Coey for reviewing earlier
drafts of this book. Thank you to Gareth Lind for his great design sense and
to everyone at Firefly for their involvement and support. This book would
not have been possible without the talent, generosity and patience of the
photographers – thanks, guys!

Hawks & Owls

OF EASTERN NORTH AMERICA

Chris G. Earley

FIREFLY BOOKS

A Firefly Book

Published by Firefly Books Ltd. 2012

Copyright © 2012 Chris G. Earley

Photographs © 2012 individual photographers as credits

Second Edition 2012

Publisher Cataloging-in-Publication Data (U.S.)

Earley, Chris.
Hawks and owls of eastern North America / Chris Earley.
2nd ed.

[144] p. : col. photos., maps ; cm.
Includes bibliographical references and index.
Summary: Profiles of numerous hawks and owls of eastern North America using color maps and photos.

ISBN-13: 978-1-55407-999-5

1. Hawks – East (U.S.). 2. Owls – East (U.S.). 3. Hawks – Canada, Eastern. 4. Owls – Canada, Eastern. I. Title.

598.944/0874 dc23   QL696.F32E37   2012

Library and Archives Canada Cataloguing in Publication

Earley, Chris G., 1968–
Hawks & owls of eastern North America / Chris Earley. – 2nd ed.

Includes bibliographical references and index.

ISBN-13: 978-1-55407-999-5

1. Birds of prey—Handbooks, manuals, etc. 2. Hawks—Canada, Eastern—Handbooks, manuals, etc. 3. Hawks—East (U.S.)—Handbooks, manuals, etc. 4. Owls—Canada, Eastern—Handbooks, manuals, etc. 5. Owls—East (U.S.)—Handbooks, manuals, etc. I. Title. II. Title: Hawks and owls of eastern North America.

QL696.F32E27 2012   598.9'44097
C2011-906743-9

Published in the United States by
Firefly Books (U.S.) Inc.
P.O. Box 1338, Ellicott Station
Buffalo, New York 14205

Published in Canada by
Firefly Books Ltd.
66 Leek Crescent
Richmond Hill, Ontario L4B 1H1

Cover and interior design by Lind Design
Front cover photo: Jim Flynn
Back cover photo: Robert McCaw
Printed in China

The publisher gratefully acknowledges the financial support for our publishing program by the Government of Canada through the Canada Book Fund as administered by the Department of Canadian Heritage.

Table of Contents

Those magnificent raptors

SEEING A HAWK RIDING LOFTY AIR CURRENTS OR HEARING an owl calling through the darkness is awe-inspiring to most observers. Yet, while many of us find these birds interesting, they can be a difficult group to identify. Many hawks zip past us too quickly, and the nocturnal habits of most owls leave us in the dark about their appearance. This book will help you identify these fascinating birds as well as learn a bit about their natural history.

Watching birds in their environment reveals interactions that link all of nature together.

When trying to identify birds it is important to remember the following motto: *I don't know.*

Really, it's okay to say it. Too many birders will get an inconclusive view of a bird and then just guess. With practice, you can identify birds from incredibly short glimpses of them, but there will always be some "I don't knows." And even if you do get a good look and still can't identify the bird, you will have learned from the process. The next time you see that species, it will be familiar to you and you may see another field mark or behavior to help in its identification. And don't forget to watch the birds as well! Keeping a checklist is fun and a way to record your sightings, but careful observations will help you really understand these interesting creatures. Watching birds in their environment reveals interactions that link all of nature together.

How to use this book

SIZE
Size is a field mark that isn't very reliable; for example, many birds often look bigger when they are flying or perched on a dead tree. But because different hawk species may migrate together and may be seen chasing or eating known prey species, size comparisons can be made. As well, both hawks and owls are often "mobbed" by small birds, offering another chance for size comparison. In this book, the raptor's length and wingspan are compared to those of a well-known species, with the measurement indicated as fairly equal to (=), less than (<) or greater than (>) the well-known bird. Average measurements for these species are given below. Remember that there are always individual variations; also, except for vultures, the females of each species described in this book are commonly larger than the males.

Length			
American Robin	10" (25 cm)	Red-tailed Hawk	19" (48 cm)
Rock Dove (Pigeon)	13" (33 cm)	Bald Eagle	31" (79 cm)
American Crow	17" (43 cm)		
Wingspan			
Mourning Dove	18" (46 cm)	Red-tailed Hawk	49" (124 cm)
Killdeer	24" (61 cm)	Herring Gull	58" (147 cm)
Rock Dove (Pigeon)	28" (71 cm)	Great Blue Heron	72" (183 cm)
American Crow	39" (99 cm)		

 = PERCHED
A description of the shape and field marks of a sitting hawk or owl.

= IN FLIGHT
A description of field marks to help identify a hawk flying overhead; unless otherwise noted, these descriptions relate to the underparts of the bird.

ADULT
Birds that have reached their adult plumage. This may be achieved by their first fall, or near their second fall, or not until four to six years of age, depending on the species.

The box at the top left of the page graphically represents the adult female hawk's or eagle's undertail coverts and tail pattern. For owls, there is a silhouette of the owl's shape and size.

FIRST YEAR

Birds that have not reached their adult plumage. Typically, most hawks have a first year plumage that lasts until the bird is approximately one year old. Eagles have additional plumages before they acquire their full adult plumage.

FLEDGLING/ FIRST FALL

Birds that have a plumage that is distinctive for a few months after they leave the nest.

MORPH

Some hawk and owl species have more than one color morph. This is not related to the sex or age of an individual or to seasonal changes; it is their color for life. This is comparable to the gray or black colors of gray squirrels or the yellow, chocolate or black colors of Labrador Retrievers.

FLIGHT TRAITS

A description of shape, behavior or flight patterns that may help identify a species.

LISTEN FOR

Learning and remembering owl calls is an important skill for identifying this group of birds, since they are more often heard than seen. Hawks also have distinctive calls, but are harder to distinguish. Only one or two examples of common calls are given here. Listening to recordings of bird calls will help you hone your recognition skills.

COMPARE TO This lists other birds that look similar to the particular hawk or owl. Comparison pages at the back of the book will be helpful when comparing similar species.

SEASONAL STATUS This list (page 17) refers to the seasonal status of hawks and owls at Point Pelee National Park. Because Point Pelee is a central point for much of the Great Lakes region, you can use this information as a guideline for when these birds may arrive or leave your area. However, because Point Pelee may not have the breeding or wintering habitat required for many species, the application of this chart during the summer and winter may be less accurate for your area.

RANGE MAPS These maps show each species' breeding and wintering ranges, as well as where they may be resident all year long.

A note to beginners

WHEN LOOKING AT HAWKS, resist the urge to instantly start flipping through this guide. Watch the bird first. This way you can look for field marks and behaviors before the bird disappears from your view. Ask yourself questions such as *Does it have pointed wings or rounded wings? Are there any markings on its breast or belly? Does it have a long or short tail? What markings does it have on its wings?*

Watch the bird for a while before flipping through this guide.

ONCE YOU'VE ANSWERED THESE AND OTHER QUESTIONS, then look in this book. A hawk flying overhead doesn't always stay in view for long and you should spend your time looking at it before it moves on.

TRY TO LEARN THE COMMON SPECIES OF YOUR AREA FIRST. Don't just observe their markings – recognize their shapes and flight patterns, too. Being familiar with the common hawks will help with identification of the less common or migratory species that you will come across.

The quotes

MANY OF THE DESCRIPTIONS in this book include a quote from naturalist writings on bird behavior and identification. While these observations may seem unscientific or "fluffy" to some readers, I believe that these naturalists have a magnificent understanding of birds and their lives. While giving nonhuman creatures human characteristics (anthropomorphism) is unscientific, I believe that beginners can benefit from this practice. What better way for a human to initially learn about something than to use humanlike descriptions? So try reading the quotes, then watch a hawk hunting or listen to an owl calling. You may find that the melodramatic or colorful style does indeed apply to your subject.

While giving nonhuman creatures human characteristics is unscientific, beginners can benefit from this practice.

Taxonomy

YOU MAY BE SURPRISED to learn that some raptor species are not very closely related (see next page). In fact, owls are more closely related to whip-poor-wills than they are to hawks. This may seem confusing, but keep in mind that many different bird groups eat other creatures. Most of the birds in this book have adapted to using their strong talons to catch their food.

Surprisingly, some raptor species are not very closely related.

The order of the birds in this book follows the seventh edition of the *American Ornithologists' Union Check-list of North American Birds*. The birds are arranged in a specific sequence (taxonomic order) that recognizes relationships between species. You may notice that many closely related birds, such as the Accipiter hawks, will have similar behaviors, shapes and flight patterns. This will hopefully help you to use shape and behavior as identification aids.

Classification of the birds in this book

HERE IS A LIST that shows how the birds in this book are classified. Each genus in a family is listed to show which species are quite closely related.

Class Aves: Birds
Order Accipitriformes: Hawks and Allies

FAMILY CATHARTIDAE: NEW WORLD VULTURES

Genus	*Coragyps*	Black Vulture
	Cathartes	Turkey Vulture

FAMILY PANDIONIDAE: OSPREY

Genus	*Pandion*	Osprey

FAMILY ACCIPITRIDAE: KITES, EAGLES, HAWKS & ALLIES

Genus	*Elanoides*	Swallow-tailed Kite
	Ictinia	Mississippi Kite
	Haliaeetus	Bald Eagle
	Circus	Northern Harrier
	Accipiter	Northern Goshawk, Sharp-shinned & Cooper's Hawks
	Buteo	Red-shouldered, Broad-winged, Swainson's, Red-tailed, Ferruginous & Rough-legged Hawks
	Aquila	Golden Eagle

Order Falconiformes: Caracaras & Falcons

FAMILY FALCONIDAE: CARACARAS & FALCONS

Genus	*Falco*	American Kestrel, Merlin, Gyrfalcon, Peregrine Falcon & Prairie Falcon

Order Strigiformes: Owls

FAMILY TYTONIDAE: BARN & BAY OWLS

Genus	*Tyto*	Barn Owl

FAMILY STRIGIDAE: TYPICAL OWLS

Genus	*Megascops*	Eastern Screech-Owl
	Bubo	Great Horned Owl, Snowy Owl
	Surnia	Northern Hawk Owl
	Strix	Barred & Great Gray Owls
	Asio	Long-eared & Short-eared Owls
	Aegolius	Boreal & Northern Saw-whet Owls

Hawk groups

HAWKS IN THE SAME GENUS share similar shapes that can be recognized in the field, allowing birders to narrow down what species they are looking at by noticing the overall shape first. Here are a few examples; others are discussed in the text for each species.

Sharp-shinned

Accipiters (Sharp-shinned, Cooper's & Northern Goshawk)
Overall these hawks have relatively short, rounded wings and long tails – adaptations for quick bursts of speed and fine maneuverability when chasing smaller birds through forested areas.

Red-tailed

Buteos (Red-tailed, Rough-legged, Broad-winged, Swainson's, Ferruginous & Red-shouldered)
These hawks have long, rounded, wide wings and relatively short tails – adaptations for soaring to look for prey. Some of these characteristics are also found in vultures and eagles.

Harrier

Northern Harrier
This species has long, rounded wings and a long tail. It coasts close to the ground while looking for small mammalian prey.

Osprey
This species has long wings that are usually held with a bend at the wrist, giving the Osprey a distinctive shape.

Osprey

Falcons (American Kestrel, Merlin, Peregrine, Prairie & Gyrfalcon)
These hawks have long, pointed wings for fast flying to chase down prey in open areas. Mississippi and Swallow-tailed Kites also have this wing shape.

Merlin

All of these hawks can modify their shape depending on whether they are *soaring* (wings out fully with completely spread wing tips and tail feathers) or *gliding* (wings somewhat bent and wing tips and tail feathers folded). Thus, a soaring Peregrine Falcon looks unfalconlike, whereas a Northern Goshawk in a glide looks quite falconlike. So, be careful when trying to put a hawk into one group or another.

Identification features

Undertail coverts

Belly band

Breast

Patagial mark

Wing lining

Wrist

Comma

Terminal band

Subterminal band

Secondaries

Primaries

Wingtips

Underparts of a flying hawk

Breast with streaking – vertical lines

WHEELER

Breast with barring – horizontal lines

FLYNN

Ear tuft   Eyebrow   Mustache   Facial disk

Owl facial features

McCAW

Primaries

Secondaries

Rump

Leading edge
of wing

Trailing
edge of
wing

Upperparts

WHEELER

15

Hawk look-alikes

OWLS ARE QUITE DISTINCTIVE due to their large heads, but hawks have a few look-alikes that sometimes confuse the observer.

Flying terns, gulls (shown) and nighthawks can sometimes look like falcons or kites because of their pointed wings, but gulls have a long beak and nighthawks have white wing patches.

The fast flight and pointed wings of the Rock Pigeon, shown here, can make it appear to be a Merlin or Peregrine to beginning birders. The Mourning Dove has a display flight that can look surprisingly similar to a gliding Sharp-shinned Hawk.

Northern (shown) and Loggerhead Shrikes are raptorial songbirds with hooked beaks that they use to kill their prey.

More than once I've had people bring me a window-killed "hawk" that ended up being a Ruffed Grouse.

Another to be aware of is the Common Raven, which is a hawk-sized "songbird" that often soars – watch for its wedge-shaped tail. American Crows, Killdeer and Purple Martins can all do a good hawk or falcon imitation, too.

Seasonal status of hawks & owls
for Point Pelee National Park

■ **Common**   ▦ **Uncommon**   — **Rare**   ····· **Very rare**

Month	J	F	M	A	M	J	J	A	S	O	N	D
Black Vulture												
Turkey Vulture												
Osprey												
Swallow-tailed Kite												
Mississippi Kite												
Bald Eagle												
Northern Harrier												
Sharp-shinned Hawk												
Cooper's Hawk												
Northern Goshawk												
Red-shouldered Hawk												
Broad-winged Hawk												
Swainson's Hawk												
Red-tailed Hawk												
Rough-legged Hawk												
Golden Eagle												
American Kestrel												
Merlin												
Gyrfalcon												
Peregrine Falcon												
Barn Owl												
Eastern Screech-Owl												
Great Horned Owl												
Snowy Owl												
Long-eared Owl												
Short-eared Owl												
Northern Saw-whet Owl												

FROM J.R. GRAHAM 1996

17

Black Vulture

Coragyps atratus

REAUME

Adult

THE BLACK VULTURE IS FOUND from the southeastern U.S. all the way through Mexico and most of Central and South America. It is a rare wanderer to the Great Lakes region. Unlike Turkey Vultures, Black Vultures can't find carrion by smell. To compensate for this, they keep a sharp eye out for descending Turkey Vultures, which can smell carcasses hidden from above by trees or shrubs. Once at the carcass, the more aggressive Black Vultures often chase the Turkey Vultures away and get to feed first.

SIZE

Length > Red-tailed Hawk
Wingspan = Herring Gull

ADULT

Naked gray head · long beak with light tip · all black body · short tail · gray legs.

FLIGHT TRAITS

Wide wings held at slight dihedral when soaring · when it flaps, the *wing beats are fast and shallow · short, triangular tail.*

ADULT

Small, naked gray head · light gray primaries contrasting with the rest of the black wings and body.

FIRST YEAR

Similar to adult except for · blackish head with no light tip on beak.

Adult   FLYNN

This partial albino Turkey Vulture has   FLYNN
a similar pattern to the Black Vulture, but
note the difference in shape.

LISTEN FOR

Hisses and grunts when interacting with other vultures at a carcass.

COMPARE TO

Turkey Vulture, Golden and Bald Eagles.

NATURE NOTES

Like other New World vultures, Black Vultures practice "urohidrosis," a big word that means they poop on their legs to stay cool. Unlike most of the birds in this book, this species is very social and family members may form groups that jointly defend roosts and carcasses from unrelated intruders.

RANGE

Resident year round

Turkey Vulture

Cathartes aura

Adult

McCAW

First fall

WHEELER

HERE IS A BIRD THAT IS OFTEN described as ugly, but just as often described as beautiful. Its red, wrinkled, naked head may be unappealing to some observers, but its flight certainly makes up for it. A Turkey Vulture gracefully wafting through the air on its upheld wings is attractive to most earth-bound humans. This species uses its nonflapping flight to slowly coast around while its exceptional sense of smell hones in on its favorite food – dead animals. When a carcass is found, the reason for the Turkey Vulture's naked head becomes apparent as soon as the bird reaches inside for the good bits. Bird heads are difficult to self-preen, so the lack of head feathers is a great adaptation utilized by many carrion-eating species. Winsor Marrett Tyler (in Bent, 1937) describes it as "a naked head and neck like the bare arms of a butcher."

SIZE

Length > Red-tailed Hawk
Wingspan < Great Blue Heron

Adult

SMALL

Note that the wings are held in a "V."

FLYNN

ADULT

Naked red head • long white-tipped beak • *blackish body* • some brown wash or edging on back and wings • long tail • pale pinkish legs.

FLIGHT TRAITS

Soars with wings in a strong dihedral or "V" • rocking or teetering flight • when it flaps, the wing beats are deep and heavy • *long tail*.

RANGE

▪ Breeding only
▪ Resident year round

McCAW

Adult

ADULT

Small, naked red head • gray flight feathers contrast with black wing linings and body.

FIRST YEAR

Similar to adult except for • some dusky coloration on the head and tip of the beak.

FLEDGLING/FIRST FALL

Similar to first year except for • all gray head and dark beak tip.

LISTEN FOR

Usually silent, but can hiss when interacting with other vultures at a carcass or when disturbed at its nest.

COMPARE TO

Black Vulture, Golden and immature Bald Eagles.

NATURE NOTES

Perched Turkey Vultures regularly spread their wings out and sun themselves. This is thought to help maintain body temperature and/or help feathers bend back into a normal shape after being bent when soaring. It may also aid in deterring parasites. One of their defence tactics when disturbed at the nest is to regurgitate, and they have also been known to "play dead."

Roosting adults

McCAW

SMALL

This individual is sunning itself, a practice that may deter parasites or help feathers keep their shape.

Osprey

Pandion haliaetus

Adult

CATCHING FISH IS THE OSPREY'S specialty, often earning it the name of Fish Hawk. They hover above a potential meal and then plunge feet-first into the water, sometimes submerging themselves in the process. An outer toe that can rotate to the back and feet that are covered in tiny spines help the Osprey to grasp a slippery fish. When it clears the water's surface and starts for a suitable feeding perch, an Osprey often holds its catch so the fish's head and tail are oriented in the same way as the bird. This reduces drag, a helpful strategy when carrying an already heavy fish. But Osprey don't always catch a fish at each attempt. Studer (1881) wrote, "Their peculiar mode of fishing necessitates the making of many a plunge to no purpose; but this does not at all discourage them: their motto always is 'try again.'"

Adult

BECK

SIZE

Length > Red-tailed Hawk
Wingspan > Herring Gull

ADULT

White head • *thick, dark brown eyeline*
• dark beak • dark brown upperparts
• *white underparts* • variable streaky dark
necklace on breast (usually more apparent on
females) • finely barred tail.

FLIGHT TRAITS

Long wings held in a downward curve
(producing a shallow "M") • *wings may bend
at the wrist (like a gull)* • may hover while
hunting.

ADULT

White underparts • possible streaked necklace
• heavily barred flight feathers, especially on

RANGE
■ Breeding only
■ Resident year round
□ Wintering only

25

Crows, jays and other birds often harrass or "mob" birds of prey.

FLYNN

secondaries • *dark wrist patches* • *dark wing tips* • finely barred tail.

FLEDGLING/FIRST FALL

Similar to adult except for • whitish edging on back and upperwings • may have buffy wash on underparts.

LISTEN FOR

High, shrill whistles and chirps.

COMPARE TO

Gulls, Bald Eagle, Rough-legged Hawk.

NATURE NOTES

The Osprey can be found over much of the planet but is absent from the Arctic and Antarctica. There are reports of Ospreys attempting to catch fish that are too big and drowning because they can't free their talons from their prey.

Osprey may totally submerge when trying to catch a fish.

FLYNN

REAUME

This osprey is trying to orient the fish's head forward to reduce drag.

REAUME

Adult with chicks

Swallow-tailed Kite

Elanoides forficatus

FLYNN

Adult

"THE KITE COURSES THROUGH THE AIR with a grace and buoyancy it would be vain to rival. By a stroke of the thin-bladed wings and a lashing of the cleft tail, its flight is swayed to this or that side in a moment ... it swoops with incredible swiftness." So wrote Coues (in Studer, 1881) and, indeed, this raptor is like a giant swallow, both in shape and elegance. It is rare in our area, but it sometimes does wander this far north, much to the delight of Great Lakes region birders.

SIZE

Length > Red-tailed Hawk
Wingspan = Red-tailed Hawk

ADULT

All white head • white underparts • shiny black back, wings and tail • very long, pointed wings and tail.

FLIGHT TRAITS

Long, pointed wings • very graceful flight • long, forked tail.

ADULT

All white head, underparts and wing linings • black back, primaries, secondaries and tail feathers.

FLEDGLING/FIRST FALL

Similar to adult except for • shorter tail.

Adult

Adult

LISTEN FOR

A high, whistled *kwee-kwee-kwee-kwee*.

NATURE NOTES

The Swallow-tailed Kite has been known to take entire small bird's nests containing young and eat the nestlings as it flies along.

RANGE

■ Breeding only
■ Resident year round

Mississippi Kite

Ictinia mississippiensis

WHEELER

Adult male

WHEELER

Fledgling to first spring

THE MISSISSIPPI KITE IS ANOTHER graceful flyer. Audubon (1840) wrote, "At times it floats in the air as if motionless, or sails in broad regular circles, when, suddenly closing its wings, it slides along to some distance, and renews its curves." The Mississippi Kite is found in the southern U.S., but is rare in the Great Lakes region. When seen here, it is usually during either spring or fall migration when some individuals may wander from their regular range.

SIZE

Length > Rock Dove
Wingspan > Rock Dove

ADULT MALE

Pale gray head • black patch in front of eye • *gray back and underparts* • black wingtips and tail • white on upperparts of secondaries.

ADULT FEMALE

Similar to adult male except for • darker gray head • whitish mottling on undertail coverts • dark gray tail with black terminal band.

SECOND YEAR

Similar to adult except for • dark gray on upperparts of secondaries • thin white bands in tail.

FLEDGLING TO FIRST SPRING

Similar to second year except for • thin dark streaks on head • white edges to feathers of upperparts • *thick rufous streaks on underparts.*

Adult male · WHEELER

Fledgling to first spring · WHEELER

FLIGHT TRAITS

Pointed wings · *overall shape falconlike* · graceful flight.

ADULT MALE

Pale gray head · *gray underparts* · *white secondaries and rufous parts in primaries seen on upperwing* · thin, white trailing edge on wings · black tail.

ADULT FEMALE

Similar to adult male except for · darker gray head · whitish mottling on undertail coverts · tail dark gray with a black terminal band.

SECOND YEAR

Similar to adult except for · dark gray secondaries and no rufous parts in primaries on upperwing · variable rufous mottling on wing linings (becomes more gray by the late part of the summer) · thin white bands in tail.

FLEDGLING TO FIRST SPRING

Similar to second year except for · *thick rufous streaks on underparts* · *thick rufous mottling on wing linings*.

LISTEN FOR

A high whistle *fee-teeew*.

COMPARE TO

Peregrine Falcon, Merlin.

RANGE	
■	Breeding only
■	Resident year round
■	Wintering only

Bald Eagle

Haliaeetus leucocephalus

Adult

SMALL

First year

WHEELER

I'M SURE ALMOST EVERYONE reading this can remember their first good look at a Bald Eagle. And if you haven't seen one yet, you will be impressed when you do. This is the largest raptor in eastern North America and, though they share the name "eagle," it is not very closely related to the Golden Eagle (pages 72–73). After a drastic decline in numbers due to shootings and the effects of pesticides, Bald Eagles are recovering and now breed in every province and state except Vermont and Rhode Island.

SIZE

Length much > Red-tailed Hawk
Wingspan > Great Blue Heron

ADULT

Large, all white head • yellow beak • yellow eyes • all dark brown body • all white tail and white undertail coverts.

Note: Figuring out the age of immature Bald Eagles is tricky due to individual variation.

THIRD & FOURTH YEAR

All dark brown with some white mottling on belly and back • pale yellow eyes • whitish head with dark mottling (may have a thick dark eyeline) • yellow beak (may have dusky patches or tip) • whitish tail with dark terminal band. (Fourth year can be very close to adult plumage.)

Adult

SMALL

🏃 SECOND YEAR

Similar to third and fourth year birds except for • darker head with less distinctive eyeline • grayish beak with dark tip • usually brown eyes.

🏃 FIRST YEAR

Similar to second year bird except for • dark brown head • less whitish mottling • buffy, light brown or dark brown belly • dark eyes.

FLIGHT TRAITS

Very long, wide wings • wings held flat or at a slight dihedral or "V" • large head • flaps with fairly stiff, slow wing beats.

🦅 ADULT

White head • brown body and wings • *white tail and undertail coverts*.

RANGE	
■	Breeding only
■	Resident year round
■	Wintering only

Third year

SMALL

THIRD & FOURTH YEAR

Whitish head with dark mottling (may have a thick, dark eyeline) • whitish tail with dark terminal band • *variable amounts of white mottling on belly, wing linings and secondaries.* (Fourth year can be very close to adult plumage.)

SECOND YEAR

Similar to third and fourth year except for • browner head • possibly more *white mottling on belly, wing linings and secondaries* • jagged trailing edge on wings (some feathers longer than others).

FIRST YEAR

Similar to second year except for • dark brown head • buffy, light brown or dark brown on belly • may have a darker tail • no jagged trailing edge on wings.

LISTEN FOR

Call described by Bent (1937) as "ridiculously weak and insignificant." A fast, chattering series of notes *ki-ki-kir-ker-ker-kur-kur.*

COMPARE TO

Golden Eagle, Turkey Vulture, Osprey.

NATURE NOTES

The oldest recorded age for a Bald Eagle in the wild is 28 years.

Second year

WHEELER

First year

Second year

Third year

Northern Harrier

Circus cyaneus

Adult male

SMALL

First year male

WHEELER

ONCE CALLED THE MARSH HAWK, the Northern Harrier is the only harrier found in North America. They have a distinctive hunting style that is well described by Thoreau (1856): "the marsh hawks flew in their usual irregular low tacking, wheeling, and circling flight, leisurely flapping and beating, now rising, now falling, in conformity with the contour of the ground." This is the only North American hawk that has facial disks, allowing it to locate prey by sound like owls can. So, by staying close to the ground, they are able to watch *and* listen effectively for prey.

Adult male in flight

WHEELER

SIZE

Length: male = American Crow
         female = Red-tailed Hawk

Wingspan: male = American Crow
          female < Red-tailed Hawk

ADULT MALE

Smoky gray head and back · *facial disks*
· yellow eyes · whitish belly with rufous spots
· dark wingtips · *white rump* (hard to see on
perched birds) · long, gray tail with dark bands
· long legs.

ADULT FEMALE

Brown head and back · *facial disks*
· yellow eyes · buffy underparts with dark
streaks · *white rump* (hard to see on perched
birds) · long, brown tail with dark bands
· long legs.

RANGE

■ Breeding only
■ Resident year round
□ Wintering only

Adult female on nest

LYNCH

🔪 FIRST YEAR

Similar to adult female except for
• unstreaked (or a few streaks), *orange-rufous breast and belly* (can fade considerably by spring) • pale eyes on males, dark brown eyes on females.

FLIGHT TRAITS

Wings held in a strong dihedral or "V"
• leading edge of wings pushed forward when soaring • long wings and tail • *often flies close to the ground* • rocks or teeters like a Turkey Vulture.

🔪 ADULT MALE

Smoky gray head and upper breast
• *white belly, undertail coverts and underwings*
• some rufous spots on belly and wing linings •
black wingtips (like a gull) • black trailing edge to secondaries

WHEELER

First year male

• whitish tail with dusky bands • *white rump when viewed from above.*

🔪 ADULT FEMALE

Brown head • buffy underparts and wing linings with dark streaks • darkish patch on wing linings close to the body • banded

Adult female

First year female showing white rump

primaries, secondaries and tail feathers • *white rump when seen from above.*

🖊 FIRST YEAR

Similar to the adult female except for
• unstreaked (or a few streaks), *orange-rufous breast, belly and wing linings* (can fade considerably by spring).

LISTEN FOR

A high *kek-kek-kek-kek-kek-kek.*

COMPARE TO

Cooper's Hawk, Northern Goshawk, Peregrine Falcon.

NATURE NOTES

The color differences between adult males and females of this species are the most pronounced of all the birds in this book (except for the American Kestrel, pages 74–77). The male's coloration is very similar to a gull's; both have smoky gray upperparts, white underparts and black wingtips. Northern Harriers nest on the ground and so are vulnerable to predators such as coyotes, foxes and skunks.

Sharp-shinned Hawk

Accipiter striatus

Adult

LITTLE BIRDS BEWARE! THE Sharp-shinned Hawk is certainly a predator that needs to be watched for. They catch their avian prey through fast, surprise attacks, as described by Studer (1881): "Its flight is peculiar – swift, spirited, and irregular, now soaring high into the air, then suddenly sweeping close to the ground. It seems to advance by sudden dashes, and when once its prey is discovered, will pounce upon it with a swiftness which makes escape impossible." Winter bird feeders often become real "bird feeders" when a local sharpie zips in and

First year

catches a junco or goldfinch. But because of their small size, these hawks have to be careful, too, or they will end up on the menu of other raptors, such as Peregrine Falcons.

Adult

WHEELER

SIZE

Length: male = American Robin
            female = Rock Dove
Wingspan: male > Mourning Dove
            female – Killdeer

ADULT

Small head • red eyes • bluish gray back, wings and tail (may have some whitish spots; female may be a brownish gray above) • *thick, rusty-orange barring on white breast and belly* • may have orange on cheeks • white undertail coverts • may have a few white spots on back • thick, dark bands on tail • *squarish tail* with thin, white terminal band (note that from below, the outer tail feathers are the same or not much shorter than others) • long, thin legs.

RANGE
- Breeding only
- Resident year round
- Wintering only

41

Adult with red eyes

FLYNN

First year with yellow eyes

McCAW

🦅 FIRST YEAR

Similar to the adult except for • yellow eyes
• brown back and wings (may have some small whitish patches) • brownish gray tail • *thick brown streaks on white breast and belly*.

FLIGHT TRAITS

Wings usually held flat when soaring
• *small head* • rounded wings • flapping of wings is floppy, not stiff • leading edge of wings often pushed forward when soaring
• *long tail with squarish tip* • *often flies with the typical accipiter pattern of a group of flaps and then a glide*.

🦅 ADULT

Red eyes • bluish gray back, wings and tail
• thick, rusty-orange barring on white breast, belly and wing linings • white undertail coverts
• dark bands on primaries, secondaries and tail feathers • banded tail with thin white terminal band.

Adult

McCAW

WHEELER

First year

FIRST YEAR

Similar to adult in flight except for • *thick, brown streaks* on white breast and belly • brown streaks on white wing linings • bands on primaries and secondaries not as distinct.

LISTEN FOR

A high *kek-kek-kek-kek-kek-kek*.

COMPARE TO

Cooper's Hawk (see comparison pages 134–37), American Kestrel, Mourning Dove.

NATURE NOTES

Sharp-shinned Hawks take prey as small as hummingbirds and as large as Ruffed Grouse, but focus on birds the size of warblers, sparrows and thrushes. Usually less than 10 percent of their diet is small mammals and insects.

Cooper's Hawk

Accipiter cooperii

Adult FLYNN

First year WHEELER

THE MID-SIZE MODEL OF OUR three accipiters, the wings and tail of a Cooper's Hawk are proportionally longer than the others. While this can be a good identification clue for experienced birders, it will take beginners a while before these shape characteristics become useful. Cooper's Hawks focus on larger prey than the Sharp-shinned; these include robins, jays, flickers, starlings and chipmunks. Even larger prey include pheasants, quail, doves, crows and gray squirrels. Another favorite, the Rock Dove or pigeon, is often hunted right above busy city streets. The next time you are walking through town and see a large, closely packed flock of pigeons flying rapidly and being reluctant to land, take a minute and see if the reason for their uneasiness is a dashing Cooper's Hawk.

SIZE

Length:	male < American Crow
	female = American Crow
Wingspan:	male = Rock Dove
	female > Rock Dove

◢ ADULT

Red eyes • bluish gray back, wings and tail (may have some whitish spots; female may be a brownish gray above) • *thick, rusty-orange barring on white breast and belly* • may have a

Adult

BECK

dark crown • white undertail coverts • thick, dark bands on tail • *rounded tail with a fairly thick white terminal band* (note that from beneath, the outer tail feathers are much shorter than the others) • long, strong-looking legs.

FIRST YEAR

Similar to perched adult except for
• yellow eyes • brown back and wings (may have some small whitish patches) • *thin brown streaks on white breast and belly* • white undertail coverts with no streaking • brownish gray tail.

FLIGHT TRAITS

Wings usually held flat when soaring
• rounded wings • *proportionally longer wings and tail than other accipiters* • large head
• straight leading edge on wings when soaring
• *wings quite stiff when flapping* • long tail is rounded at the tip • *often flies with the typical accipiter pattern of a group of flaps and then a glide.*

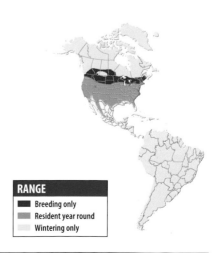

RANGE
- Breeding only
- Resident year round
- Wintering only

45

Adult

WHEELER

ADULT

Thick, rusty-orange barring on white breast, belly and wing linings • white undertail coverts • dark bands on primaries, secondaries and tail feathers • banded tail with fairly thick white terminal band.

FIRST YEAR

Similar to adult in flight except for • narrow brown streaks on white breast and belly • brown streaks on white wing linings • bands on primaries and secondaries not as distinct.

LISTEN FOR

A loud alarm call of *kak-kak-kak-kak-kak-kak.*

COMPARE TO

Sharp-shinned Hawk, Northern Goshawk (see comparison pages, 134–37).

NATURE NOTES

It is thought that one of the main food items for this species was once the now extinct Passenger Pigeon.

This shows the extreme size difference between male and female accipiters.

First year

Northern Goshawk

Accipiter gentilis

Adult   FLYNN

First year   LYNCH

"BEAUTY AND COURAGE, SWIFTNESS and strength mean something to us; and we shall find these qualities in high degree in the hawks of the Accipiter clan. Especially is this true of the largest and strongest of them, the goshawk, one of the deadliest, handsomest, bravest birds of prey in the world," wrote Sass (1930, in Bent). If you've ever been near a Northern Goshawk's active nest, then this quote probably hits home. And "hitting home" is exactly what the parents will try to do if you are close to their nest. Their loud "kak, kak, kak, kak, kak" alarm call gives you a short warning that ducking your head might be a good idea. Just thinking about it makes my spine tingle. The Northern Goshawk is our largest accipiter and it focuses on prey such as hares, squirrels, grouse, woodpeckers, jays, crows and smaller birds. I used to watch one hunt on a farm that had a bird feeder on one side of the farmhouse and a driveway on the other. The goshawk would fly rapidly and low up the driveway, then swoop around one end of the house and suddenly appear at the feeder, much to the dismay of the local Mourning Doves.

Adult

WHEELER

SIZE

Length: male = Red-tailed Hawk
female > Red-tailed Hawk
Wingspan: male = American Crow
female > American Crow

🖋 ADULT

Dark crown and eyeline • *thick white eyebrow*
• red eyes • bluish gray back, wings and tail •
very thin, gray barring on white breast and belly
(can be thicker on females) • white undertail
coverts • strong-looking legs.

🖋 FIRST YEAR

Similar to perched adult except for • yellow
eyes • thinner eyebrow and no eyeline •
brown back and wings (may have some small
whitish patches) • *thick brown streaks on buffy
or whitish breast and belly* • some streaks on
white undertail coverts • brownish gray tail

RANGE
Resident year round
Wintering only

First year

WHEELER

Very thin, gray barring on white breast, belly and wing linings (can be thicker on females) • white undertail coverts • dark bands on primaries • upperwing has dark primaries and secondaries with the rest of the wing lighter gray.

FIRST YEAR

Similar to adult in flight except for • *thick brown streaks on buffy or white breast and belly* • *some brown streaks on undertail coverts* • brown streaks on white wing linings • bands on primaries and secondaries • bands on tail appear jagged, especially from above • each tail band has a thin, light border (if you are close enough to see it) • pale light bar on upperwing.

LISTEN FOR

A loud alarm call of *kak-kak-kak-kak-kak-kak*.

COMPARE TO

Cooper's Hawk (see comparison pages, 134–37), Red-shouldered Hawk, Red-tailed Hawk, Gyrfalcon.

NATURE NOTES

Northern Goshawks have been known to kill other raptor species such as Sharp-shinned Hawks, Long-eared Owls, Red-tailed Hawks and even other goshawks.

with dark, thick, jagged bands • each tail band has a thin, light border on upper surface.

FLIGHT TRAITS

Rounded wings *but often have a tapered or pointed appearance* • *most buteolike accipiter* • large head • straight leading edge on wings when soaring • flaps deeply, like a buteo • long, broad tail • *often flies with the typical accipiter pattern of a group of flaps and then a glide.*

First year

First year

Buteo lineatus

Adult

McCAW

THIS BUTEO PREFERS WET FORESTS and wooded river valleys. Here it often hunts from a perch, watching for frogs, small mammals, snakes, small birds, insects, fish, earthworms and crayfish. In other words, the Red-shouldered Hawk isn't too choosy. It used to be much more common in the Great Lakes region, but habitat degradation has caused its numbers to decline here.

SIZE

Length = American Crow
Wingspan = American Crow

ADULT

Brown head and back with some whitish mottling • *blackish primaries, secondaries and tail feathers with thin white bands* • *reddish shoulders* (not always apparent) • thick, rusty-orange barring on white breast and belly • white undertail coverts • short tail (note: birds in Florida are much paler overall).

FIRST YEAR

Similar to adult except for • brown primaries, secondaries and tail feathers with dark bands • thick brown streaks on white breast and belly.

FLIGHT TRAITS

Wide, rounded wings • stiff wing beats • wings held flat or droop slightly when soaring • broad tail.

ADULT

Rusty-orange barring on white breast and belly • *rusty-orange wing linings* (may be paler than breast and belly) • thin black

First year

WHEELER

HOLDEN

Adult – note "windows" on wings

WHEELER

First year – note "windows" on wings

bands on primaries and secondaries • dark tips on primaries • *translucent "window" near tips of primaries* (note: other hawks may have "windows" in their wings while they are molting; these are different than the "windows" of a Red-shouldered, but can be confusing for beginning birders) • thick black trailing edge on secondaries • white undertail coverts • *black tail with two or three thin white bands visible plus a thin white terminal band.*

FIRST YEAR

Similar to adult except for • thick dark streaks on white breast and belly • white wing linings • indistinct bands on primaries and secondaries • indistinct dark trailing edge on secondaries • dark tips on primaries • tail with many dark bands.

LISTEN FOR

A loud, screamed *keeyaah.*

COMPARE TO

Broad-winged Hawk, Red-tailed Hawk, Northern Goshawk.

NATURE NOTES

The Red-shouldered Hawk has an elaborate and noisy courtship display flight that includes fast dives and climbs while giving loud screams.

RANGE
- Breeding only
- Resident year round
- Wintering only

Broad-winged Hawk

Buteo platypterus

WHEELER

Adult light morph

WHEELER

First year

OUR SMALLEST BUTEO, THE Broad-winged Hawk is another raptor who opts for a wide variety of prey. They eat small mammals, young birds, insects and seem to have a liking for frogs and toads as well. Brewster (1925, in Bent), wrote, "After alighting on a low branch or stub overlooking some shallow reach of calm water besprinkled with innumerable floating toads absorbed in the cares and pleasures of procreation, and rending the still air with the ceaseless din of their tremulous voices, the Hawk will often gaze down at them long and listlessly, as if undecided which particular one to select from among so many, or dreamily gloat over the wealth of opportunities for such selection."

SIZE

Length < American Crow
Wingspan < American Crow

ADULT

Brown head and back • no distinct bands on primaries and secondaries • thick, rusty-orange barring on white breast and belly • white undertail coverts • *short tail with thick white band*.

Adult light morph

HOLDEN

FIRST YEAR

Similar to perched adult except for • thick dark streaks on white breast and belly
• short tail with one dark, thick subterminal band and many thin bands.

FLIGHT TRAITS

Wings held flat when soaring • slightly pointed, wide wings • stiff wing beats
• fairly short tail.

ADULT

Rusty barring on white breast and belly
• white wing linings with a few rusty bars
• black primary tips • thick, black trailing edge on wings • tail has a thin, white terminal band, then two thick black bands separated by one thick white band.

RANGE

■ Breeding only
■ Resident year round
□ Wintering only

First year dark morph   WHEELER

Adult dark morph   WHEELER

FIRST YEAR

Similar to adult in flight except for • dark streaks on white breast and belly • white wing linings with variable dark streaks • dusky primary tips • thick, dusky trailing edge on wings • tail has a white terminal band, a fairly thick, dark subterminal band and many thin, dark bands.

DARK MORPH

The dark Broad-winged Hawk is a rare western-range morph. It is very rarely seen in the east during migration. This morph has primaries, secondaries and *tail feathers* similar to the light morph, but the body and the wing linings are very dark brown to black.

LISTEN FOR

A high, thin, two-parted whistle with the second part slightly lower than the first.
A *peeyurr*.

COMPARE TO

Red-shouldered Hawk, Red-tailed Hawk.

NATURE NOTES

The Broad-winged Hawk gives us one of the most spectacular migration displays in the animal kingdom. Hundreds of these hawks may circle together in a flock or "kettle" as they use uprisings of warm air called thermals to stay aloft. On September 17, 1999, in the Lake Erie Metropark, the Southern Michigan Raptor Research group counted an amazing 555,371 Broad-winged Hawks migrating southwards.

Migrating Broad-winged Hawks in a "kettle"

WHEELER

First year

WHEELER

Swainson's Hawk

Buteo swainsoni

Adult light morph

WHEELER

First year light morph

WHEELER

ANOTHER BUTEO THAT MIGRATES in large flocks, the Swainson's Hawk is a prairie bird of the west that is sometimes found in the Great Lakes region during migration. It winters in central and southern South America – a distance of over 6,250 miles (10,000 km) from its breeding range in Canada. This means it travels over 12,500 miles (20,000 km) round-trip each year. Unfortunately, they are in much danger while in many parts of their winter range where thousands die every year because of pesticide poisoning.

Adult light morph

WHEELER

SIZE

Length = Red-tailed Hawk
Wingspan = Red-tailed Hawk

🦅 LIGHT MORPH ADULT

Dark head with white forehead and throat •
white underparts with solid rufous breast • dark
brown back and wings
• short tail with thick, dark subterminal band
and many thinner bands.

🦅 LIGHT MORPH FIRST & SECOND YEAR

Buffy or whitish head (especially on early
spring birds) • dark brown back and wings with
white mottling • whitish underparts with dark
streaks (*often forming a dark patch on the sides
of the breast*) • tail with many thin, dark bands.

RANGE	
■	Breeding only
▨	Wintering only

Adult intermediate morph

WHEELER

Adult dark morph

WHEELER

First year dark morph

WHEELER

FLIGHT TRAITS

Wings held in a dihedral or "V" when soaring • rocks or teeters like a Turkey Vulture or Northern Harrier • long, wide wings that are quite pointed for a buteo • relatively long, broad tail.

LIGHT MORPH ADULT

White throat • *white underparts with a solid rufous breast* • *white wing linings contrasting with the dark primaries and secondaries* • finely barred tail with a thick, dark subterminal band.

INTERMEDIATE MORPH ADULT

Similar to light morph adult except for • dark brown breast • rufous belly • rufous in wing linings.

DARK MORPH ADULT

Similar to intermediate morph adult except for • no white throat • dark brown to black breast

Adult intermediate morph

WHEELER

Adult dark morph

WHEELER

and belly • white undertail coverts (may have dark marks) • rufous or dark markings on wing linings.

LIGHT MORPH FIRST YEAR

Similar to light morph adult in flight except for • *solid patches on sides of breast only* • dark streaks on breast and belly • dark streaks on wing linings • primaries, secondaries and tail feathers not as dark • less distinct thick subterminal band on tail. (Second year birds similar except with mostly adult primaries, secondaries and tail feathers.)

INTERMEDIATE & DARK MORPH FIRST YEARS

Similar to light morph first year except for • dark mottling on breast and belly • dark mottling in wing linings.

LISTEN FOR

A shrill, screaming whistle *kweeeeeeee*.

COMPARE TO

Dark morph to dark morph Rough-legged Hawk and Harlan's Red-tailed Hawk.

NATURE NOTES

These hawks can be fierce defenders of their nests. As a young, budding 17-year-old naturalist, I was excited to find a Swainson's Hawk nest on my uncle's farm in southern Saskatchewan. While looking up at the chicks, a rushing sound preceded one of the parent's knocking my cap off. As it dove again and again, I ran terrified to my motorcycle and took off. That didn't deter the bird at all, and it almost succeeded in knocking me off the racing motorcycle!

Red-tailed Hawk

Buteo jamaicensis

Adult Eastern

McCAW

First year Eastern

SMALL

THE RED-TAILED HAWK IS LIKELY the best known raptor in this book and one of the easiest of our raptors to identify – at least the adults are. Try not to focus too much on the red tail for identification purposes, though. Learn other field marks and behaviors so you will have a better "comparison hawk" when looking at other species. The relative abundance of Red-tailed Hawks also allows birders to observe much of their breeding biology, from courtship to nest building to raising their young.

SIZE

Length > American Crow
Wingspan much > American Crow

EASTERN ADULT

Brown head • dark eyes • white throat • brown back and wings • *light marks on wings form a "V" that can be seen from quite a distance* • whitish or buffy underparts with variable amounts of dark streaks on belly (*often forming a belly band*) • *short, red tail.*

EASTERN FIRST YEAR

Similar to adult perched except for • yellow eyes • whiter underparts with more dark streaking • *brown tail with many thin bands.*

FLIGHT TRAITS

Wings held in a very shallow dihedral or "V" when soaring • wide, long, rounded wings • fluid wing beats • broad tail.

WHEELER

Adult Eastern (a fairly pale individual – the photo at the bottom of page 65 is more typical for this area)

EASTERN ADULT

Whitish or buffy underparts with variable amounts of dark streaks on belly (*often forming a belly band*) • white or buffy wing linings • *dark patagial mark* • *dark "commas" on wrists* • dark tipped primaries • dark trailing edge to secondaries • banded primaries and secondaries • *short, red tail*.

EASTERN FIRST YEAR

Similar to adult in flight except for
• yellow eyes • whiter underparts with more dark streaking • *upperwing has light brown outer half and darker brown inner half* • *tail is brown with many thin bands*.

RANGE	
■	Breeding only
■	Resident year round

Adult

McCAW

MORPHS

There is a lot of variation in the Red-tailed Hawk across North America, especially in the west. Many look similar to our eastern birds, but there are some dark and intermediate morphs. Still others are quite a bit paler than the eastern birds. For identification purposes of flying birds, all Red-tailed Hawks have *dark patagial marks* (except for dark morphs, where the wing linings are all dark). As well, all juvenile Red-tailed Hawks have narrow bands on their tails. All adults have a *red or reddish tail*, except for one, the Harlan's Red-tail, which has a whitish to gray tail that blends to a darker gray on the tip. Check out the photos to see some of these variations, but keep in mind that they are all rare in the Great Lakes region.

LISTEN FOR

A rough, screamed descending whistle *keeee-arrrrrrr*.

COMPARE TO

Rough-legged Hawk, Ferruginous Hawk.

NATURE NOTES

Partial albinism seems to occur more frequently in Red-tailed Hawks than in other raptor species. These birds may have white feathers in patches or covering most of the body.

First year Eastern

WHEELER

First year Eastern

WHEELER

Adult Harlan's

WHEELER

Adult dark morph

WHEELER

Adult showing patagial mark and belly band

McCAW

Ferruginous Hawk

Buteo regalis

McCAW

Adult light morph and adult dark morph

THIS PRAIRIE BUTEO IS SO ROBUST that some think it should be called an eagle. Its head and beak shape look surprisingly eaglelike as well. It is a very rare visitor to most of the Great Lakes region during migration, but is regular in Minnesota.

SIZE

Length > Red-tailed Hawk
Wingspan > Red-tailed Hawk

🔾 LIGHT MORPH ADULT

Large head and beak • grayish head with white throat • rufous back and wings with gray primaries and secondaries • *white underparts with some rufous barring on belly sides and leg feathers* • *whitish or light rufous tail.*

🔾 LIGHT MORPH FIRST YEAR

Similar to adult except for • yellow eyes
• brown back and wings • some dark streaking on belly and legs, no rufous
• brownish tail with a few narrow bars.

Adult light morph
WHEELER

Adult dark morph
WHEELER

FLIGHT TRAITS

The most eaglelike buteo • large head • wings held in a dihedral or "V" when soaring • long, wide, rounded wings • flaps with fluid wing beats • long, broad tail.

LIGHT MORPH ADULT

White underparts • rufous leg feathers • whitish or light rufous tail • white wings with some rufous barring on wing linings • dark wing tips • the base of the primaries on the upperwing forms a light patch.

LIGHT MORPH FIRST YEAR

Similar to adult in flight except for • no rufous on leg feathers • no rufous on wing linings, dark spots instead • faint narrow bars on tail.

DARK MORPH

All dark brown body and wing linings • *narrow white comma near wrist in flight* • white primaries, secondaries and tail feathers with some very faint barring.

LISTEN FOR

A weak, low whistle *keeeer*.

COMPARE TO

Rough-legged Hawk, Red-tailed Hawk.

NATURE NOTES

When bison still roamed the prairies, their bones were often used as part of the Ferruginous Hawk's nesting material.

RANGE
- Breeding only
- Resident year round
- Wintering only

Rough-legged Hawk

Buteo lagopus

WHEELER

Adult male light morph

A WINTER VISITOR TO THIS AREA, the Rough-legged Hawk is a strikingly handsome buteo. Bent (1937) wrote, "the mention of its name brings visions of a splendid bird, one of the largest and finest of our hawks. Past master in the use of air currents, whether it is posed motionless in a breeze over a cliff, or scaling close to the ground and quartering like a harrier, or swinging proudly in great circles up and up into the blue sky, this great hawk is always a thing of beauty." The sun's reflections on a field of newly fallen snow often light up the undersides of flying Rough-legs, making their markings even more appealing. The dark morph of this hawk is the only commonly seen dark morph hawk in the Great Lakes region.

First year light morph

WHEELER

Adult light morph

WHEELER

SIZE

Length > Red-tailed Hawk
Wingspan > Red-tailed Hawk

🔰 LIGHT MORPH ADULT MALE

Pale head • dark eyes • dark upper breast with white mottling • *white lower breast and belly with dark mottling* • white undertail coverts • dark back with white mottling • *light tail with white terminal band, thick, dark subterminal band and one or two narrow, dark bands.* (Note: there is a lot of variation between males and females, so not all birds can be sexed by plumage.)

🔰 LIGHT MORPH ADULT FEMALE

Similar to the light morph adult male except for • light upper breast with dark streaks • *more solidly dark belly band* • *whitish tail with white terminal band, thick, dark subterminal band and a thick, dusky band.*

RANGE

■ Breeding only
■ Wintering only

69

Adult male dark morph   WHEELER

Adult female light morph   WHEELER

LIGHT MORPH FIRST YEAR

Similar to light morph adult female except for
• light eyes • only the thick, dusky subterminal
band on tail.

DARK MORPH

Very dark brown head, wing linings and body
(male may be black overall) • dark eyes (first
year has light eyes) • *light tail with white
terminal band, thick, dark subterminal band and
zero to three narrow, dark bands.*

FLIGHT TRAITS

*Wings held in a dihedral or "V," though the outer
parts of the wings are quite horizontal* • long,
wide, rounded wings • flaps with fluid wing
beats • relatively long, broad tail.

LIGHT MORPH ADULT MALE

Pale head • dark eyes • dark upper breast with
white mottling • *white lower breast and belly
with dark mottling* • white undertail coverts
• light wing linings with dark mottling •
dark wrist patch • dark tips to primaries and
secondaries • some light banding in secondaries
• *light tail with white terminal band, thick, dark
subterminal band and one or two narrow, dark
bands.* (There is a lot of variation between
males and females, so not all birds can be sexed
by plumage.)

LIGHT MORPH ADULT FEMALE

Similar to the light morph adult male except
for • light upper breast with dark streaks • *more
solidly dark belly band* • brown mottling in wing
linings • *darker wrist patches* • whitish tail with
white terminal band, thick, dark subterminal
band and a thick, dusky band.

Adult dark morph WHEELER

First year dark morph WHEELER

🔪 LIGHT MORPH FIRST YEAR

Similar to light morph female except for
• light eyes • less defined dark tips to
secondaries • less brown in wing linings • only
the thick, dusky subterminal band on tail.

🔪 DARK MORPH

Very dark brown head, wing linings and body
(male may be black overall) • dark eyes (first
year has light eyes) • white primaries and
secondaries • dark tips on primaries • *dark
wrist patches show on brown individuals* • dark
banding in secondaries • *light tail with white
terminal band, thick, dark subterminal band and
zero to three narrow, dark bands*.

First year light morph WHEELER

LISTEN FOR

A thin whistled *keeeyurrrrr*.

COMPARE TO

Ferruginous Hawk, Red-tailed Hawk, Osprey.

NATURE NOTES

Compared to other buteos of similar size, the
Rough-legged Hawk has proportionally smaller
feet and beak. This relates to its preference for
smaller prey, though it does take larger prey
occasionally.

Golden Eagle

Aquila chrysaetos

Adult

"THIS MAGNIFICENT EAGLE HAS long been named the King of Birds and it well deserves the title. It is majestic in flight, regal in appearance, dignified in manner and crowned with a shower of golden hackles about its royal head," (Bent, 1937). It's hard to think of many birds more awe-inspiring than the Golden Eagle. A formidable predator, there are even records of these eagles working together and attacking full-grown pronghorn antelope and deer! But very large prey are rarely taken and the Golden Eagle's regular diet includes hares, marmots and carrion.

SIZE

Length = Bald Eagle
Wingspan > Great Blue Heron

ADULT

Dark brown head, body and wings
• *tawny nape and eyebrow* • tawny-brown undertail coverts • dark tail with narrow lighter bands.

FIRST & UP TO FIFTH YEAR

Similar to adult except for • second to fifth years have variable whitish mottled base to tail • *first year has all white base to tail, leaving a thick, dark terminal band.*

FLIGHT TRAITS

Wings held flat or in a shallow dihedral or "V" • *relatively small head* • long, wide, rounded wings • flaps with slow, fluid wing beats.

Adult

WHEELER

First year

EGRESSY

▲ ADULT

Dark brown head, wing linings and body
• tawny nape and eyebrow • tawny-brown
undertail coverts • dark primaries, secondaries
and tail feathers with faint banding.

▲ FIRST & UP TO FIFTH YEAR

Similar to adult except for • second to fifth
years have variable whitish mottled base on
tail and may have some white bases to outer
secondaries and inner primaries • *first year
has all white base to tail, leaving a thick, dark
terminal band • first year usually shows white
patch at the base of the outer secondaries and
inner primaries.*

"Mantling" over its kill

LYNCH

LISTEN FOR

Shrill yelps.

COMPARE TO

Immature Bald Eagle, Turkey Vulture.

RANGE	
■	Breeding only
■	Resident year round
■	Wintering only

American Kestrel

Falco sparverius

Adult male

FLYNN

Adult female

SMALL

THIS IS THE SMALLEST OF OUR falcons and though it is less powerful than the others, it makes up for this by being the most colorful hawk in the world. It also has a special hunting style that is not frequently used by other hawks. Winsor Marrett Tyler wrote, "Perhaps the most remarkable of its aerial accomplishments, the bird, arresting its flight through the air, hovers, facing the wind, its body tilted upward to a slight angle with the ground, its wings beating lightly and easily." From its held position, the kestrel is able to drop onto an unsuspecting meadow vole or grasshopper. It also hunts from perches and is often seen sitting on telephone wires where it may be mistaken for a Mourning Dove.

SIZE

Length = American Robin
Wingspan < Killdeer

ADULT MALE

Bluish gray cap with rufous central spot
• *white face with thick black mustache and sideburns* • two large black spots on back of head • *rufous back with dark barring on lower back* • *bluish gray wings with dark spots* • pale rufous breast blending to whiter undertail coverts • dark spots on belly • *long, narrow, bright rufous tail*

Adult male

WHEELER

• *white outer tail feathers, which may have dark spots* • *tail has thin, white terminal band and thick, black subterminal band*
• *often pumps its tail*.

ADULT FEMALE

Similar to adult male except for • more subdued head pattern with gray cap • white underparts with rufous streaks on breast and belly • wings rufous with dark barring • *tail rufous with narrow dark bands* (subterminal band thicker than others).

FIRST YEAR MALE

Similar to adult male except for • more dark barring on back • thin dark breast and belly streaks (first year females very similar to adult females).

FLIGHT TRAITS

Long, pointed wings • wings held flat
• *long, narrow tail* • *not a strong-looking flyer* like the other falcons • *often hovers*.

RANGE

■ Breeding only
■ Resident year round
□ Wintering only

McCAW

Male Kestrel with a baby turtle it has just caught

ADULT MALE

White throat and black mustache • *pale rufous breast blending to whiter undertail coverts* • dark spots on belly • pale rufous wing linings with black spots • primaries and secondaries with many thin, dark bands • *long, bright rufous tail with thin, white terminal band and thick, black subterminal band* • white outer tail feathers with variable amounts of dark spots.

ADULT FEMALE

Similar to adult male except for • white underparts with rufous streaks on breast and belly • light wing linings with light rufous spots • *long tail rufous with narrow dark bands* (subterminal one thicker than others).

FIRST YEAR MALE

Similar to adult male except for • thin dark breast and belly streaks (first year females very similar to adult females).

LISTEN FOR

A loud *killy-killy-killy-killy-killy*.

COMPARE TO

Merlin, Sharp-shinned Hawk, Mourning Dove.

NATURE NOTES

The American Kestrel often nests in old woodpecker holes and will also use nest boxes (see page 111). Because kestrels can see ultraviolet light, and because vole urine reflects that color, it's possible that kestrels can follow vole urine trails to potential prey.

Adult female

Adult female at nest cavity in abandoned ranch building

Merlin

Falco columbarius

Adult male

First year female

THE MERLIN MAY BE LITTLE, BUT to small birds it is an unforgiving predator. Merlins often hunt by perching near an open field or lake and waiting for a small bird to cross. Then the falcon just flies down its meal with an impressive burst of speed. They also catch flying insects such as dragonflies, as well as bats that are leaving their daytime cave roosts.

SIZE

Length = American Robin
Wingspan = Killdeer

Adult male

HOLDEN

Note the thin white bands in the tail

SMALL

ADULT MALE

Bluish gray head and upperparts • dusky, indistinct mustache • pale rufous to whitish underparts with dark streaks • *black tail with light narrow bands*.

ADULT FEMALE & FIRST YEAR

Similar to male except for • whiter underparts with dark streaks • brown upperparts.

FLIGHT TRAITS

A very strong flyer • relatively short, pointed wings.

ADULT MALE

Dusky, indistinct mustache • whitish throat • pale rufous to whitish underparts with dark streaks • dark underwings with white spotting • *black tail with narrow white bands*.

ADULT FEMALE & FIRST YEAR

Similar to adult male except for • whiter underparts with dark streaks.

WESTERN MERLINS

Merlins in the prairies are lighter overall and those on the west coast can be very dark.

LISTEN FOR

A loud *kek-kek-kek-kek-kek*.

COMPARE TO

American Kestrel, Peregrine Falcon, Rock Dove.

NATURE NOTES

Merlins may cache excess prey in conifer branches or old crow nests and return to it later.

RANGE
- ■ Breeding only
- ■ Resident year round
- ■ Wintering only

Gyrfalcon

Falco rusticolus

White morph

BECK

THE GYRFALCON IS THE LARGEST falcon in the world and also the most northerly breeder, nesting as far as 82°N in Greenland. In winter, it may visit the Great Lakes region where it replaces its regular summer diet of ptarmigan with that of waterfowl and Rock Doves.

SIZE

Length > Red-tailed Hawk
Wingspan = Red-tailed Hawk

GRAY MORPH ADULT

Gray head with white throat • gray back and wings with faint barring • white underparts with gray barring • gray tail with narrow bands.

WHITE MORPH ADULT

All white head, body and tail • dark gray spots on back and wings • may have some dark barring on belly • white tail with dark, narrow bands.

DARK MORPH ADULT

Dark brown head with whitish throat • dark brown body and wings • dark brown underparts with whitish streaks • brown tail with dark, narrow bands.

FLIGHT TRAITS

Long, pointed wings, but rounder tips than other falcons • long tail • *strong flight*.

GRAY MORPH ADULT

White underparts and wings with gray barring • gray tail with dark, narrow bands.

WHITE MORPH ADULT

All white underneath • may have dark barring on belly • black wingtips • tail with narrow bands.

Gray morph

SMALL

White morph

FAIRBAIRN

Gray morph

BECK

Dark morph

BECK

🔪 DARK MORPH ADULT

Dark brown underparts and wing linings with whitish streaks • dark primaries and secondaries • dark brown tail with narrow bands.

🦅 FIRST YEAR MORPHS

Similar to adults but more streaked underparts and darker upperparts.

LISTEN FOR

Like most other falcons, a *kak-kak-kak-kak-kak*.

COMPARE TO

Peregrine Falcon, Northern Goshawk.

RANGE
- ■ Breeding only
- ■ Wintering only

Peregrine Falcon

Falco peregrinus

Adult

HOLDEN

First year

FLYNN

IT HAS BEEN ESTIMATED THAT THIS falcon may reach speeds of over 200 miles per hour (320 km/h) when it is swooping or diving for prey. It also catches flying prey by level chases or by surprise attacks. The Peregrine Falcon is one of the most widely distributed birds in the world, being found across all continents except Antarctica. Because of this immense range, its prey list likely covers over 1,000 species of birds, ranging from hummingbirds and songbirds to cranes and geese. Peregrines may also take small mammals, insects and fish occasionally.

Adult

EGRESSY

SIZE

Length = American Crow
Wingspan > American Crow

ADULT

Black cap • *black mustache* • white cheek and throat • dark gray back and wings • wing tips almost reach the tip of the tail • white underparts with dark barring on belly and undertail coverts • gray tail with dark, narrow bands.

FIRST YEAR

Buffy crown • *dark mustache* • dark patch around eye • brown back • whitish or buffy underparts with dark streaks • brown tail with dark, narrow bands.

FLIGHT TRAITS

Very strong, fluid flight • when flapping, it uses a "rowing" motion • *long, very pointed wings* • long tail • soars with wings held flat.

ADULT

Black mustache • white throat and breast • white belly, undertail coverts and wing linings with dark barring • gray tail with dark, narrow bands.

FIRST YEAR

Dark mustache • whitish or buffy throat and breast • whitish or buffy underparts and wing linings with dark streaks on belly and wing linings • brown tail with narrow bands.

WESTERN

First years of western Peregrines are much darker than the above description.

LISTEN FOR

A loud *kak-kak-kak-kak-kak-kak-kak*.

COMPARE TO

Gyrfalcon, Merlin, Northern Goshawk, Mississippi Kite, Rock Dove.

NATURE NOTES

There are records of Peregrines hunting at night. Due to the efforts of many conservation groups, Peregrines are becoming more common in the Great Lakes region. Many large cities now support nesting pairs.

RANGE
■ Breeding only
■ Resident year round
□ Wintering only

Prairie Falcon

Falco mexicanus

Adult

LYNCH

First year

WHEELER

THIS WESTERN FALCON WILL HUNT from perches or fly quickly close to the ground and surprise ground squirrels or small birds such as Horned Larks and Western Meadowlarks. The Prairie Falcon is rare in most of the Great Lakes region, but is considered regular in Minnesota.

SIZE

Length = American Crow
Wingspan = American Crow

ADULT

Thin, white eyebrow • white patch behind eye • brown head and mustache • brown back and wings • white underparts with dark speckles on breast and belly • light brown tail with faint, narrow barring.

Note the black "armpits"

WHEELER

FLIGHT TRAITS

Wings pointed, but less so than Peregrine's • strong and swift flyer.

ADULT

White throat, breast and belly • dark speckles on breast and belly • *light primaries, secondaries and tail feathers with faint barring* • black "armpits" and parts of the wing linings.

FIRST YEAR

Similar to adult except for • dark streaks on breast and belly.

LISTEN FOR

A shrill *kek-kek-kek-kek-kek*.

COMPARE TO

Merlin, Peregrine Falcon, Gyrfalcon.

RANGE

- ▮ Breeding only
- ▮ Resident year round
- ▯ Wintering only

Barn Owl

Tyto alba

Adult

SMALL

Adult in flight

McCAW

THE BARN OWL IS RESPONSIBLE FOR many folk tales that include evil owls. Because they often nest in attics and church steeples, the human inhabitants would often hear the owls' screams and hisses late at night. It is no wonder that these frightened people thought ill of them. Yet, Barn Owls have done their share to help humans as well. It has been estimated that a 10-year-old Barn Owl would have eaten approximately 11,000 mice in its lifetime. And, calculating the amount of food these mice would have eaten if they had survived for one year, this one Barn Owl would have saved about 13,000 tons of grain or crops! No wonder they are encouraged to nest in hay lofts in many areas. Barn Owls were once much more common in the Great Lakes region, but numbers have dropped dramatically in many parts of their northern range.

SIZE

Length = American Crow
Wingspan > American Crow

ADULT

White, heart-shaped face • *dark eyes* • **buffy-gold back with gray mottling and small white and black spots** • *white underparts,*

Adult WHEELER

Adult FLYNN

variably washed with buffy-gold and small black spots (females are usually darker overall) • short tail • *long legs*.

ADULT

Very pale underwings, giving the bird an overall white appearance from below.

LISTEN FOR

Blood-curdling screams, hisses, purrs, snores and twitters.

COMPARE TO

Snowy Owl.

NATURE NOTES

This owl is one of the most accurate at finding prey by sound. It has been given many other names, including White Owl, Golden Owl, Church Owl, Ghost Owl, Hissing Owl, Monkey-faced Owl, Demon Owl and Phantom Owl.

RANGE
Resident year round

Gray morph

SMALL

Red morph

McCAW

"WISE MIDNIGHT HAGS! IT IS NO honest and blunt *tu-whit to-who* of the poets, but, without jesting, a most solemn graveyard ditty, the mutual consolations of suicide lovers remembering pangs and the delights of supernal love in the infernal groves. And yet I love to hear their wailing, their doleful responses, trilled along the woodside…" Thoreau (1845) certainly summed up the eerie sounds of the screech-owl, but also described them as "a loud, piercing scream, much like the whinner of a colt perchance, a rapid trill, then subdued or smothered a note or two." Eastern Screech-Owls are one of the most familiar of owls of the southern Great Lakes region and eastern North America. Check out page 111 to see how you may be able to attract one to your backyard.

SIZE

Length < American Robin
Wingspan > Mourning Dove

▶ GRAY MORPH ADULT

Ear tufts (can be flattened and hard to see, but are usually quite prominent) • yellow eyes • thin, black border to facial disks • white eyebrows and mustache • pale yellowish green beak • *gray upperparts with dark mottling and a line of white spots on back* • some white spots

Gray morph

McCAW

EGRESSY

Brown morph – note that ear tufts may be laid back on head.

on wings • white underparts with thin dark barring and streaking • short tail.

RED MORPH ADULT

Similar to gray morph except for • gray replaced by *rich rufous*.

BROWN MORPH ADULT

An intermediate of the first two morphs, it is similar to gray morph except for
• gray replaced by *brown*.

ADULT

Pale underwings with a black comma at the wrist.

LISTEN FOR

Two main calls. The tremolo is a modulating whistle that a pair uses to stay in contact with each other. The shriller, falling whinny is a territorial call.

COMPARE TO

Long-eared Owl, Northern Saw-whet Owl.

NATURE NOTES

For their size, Eastern Screech-Owls are fierce little predators, taking a variety of prey, including birds much larger than themselves. I watched one fly from its nesting cavity to a perch and cough up a pellet (a ball of indigestible material including bones, hair and feathers). After it left, I examined the ground below this perch and found not only more pellets, but also fish bones and crayfish claws, showing their knack for catching whatever is available.

RANGE

Resident year round

Great Horned Owl

Bubo virginianus

Adult with skunk

McCAW

Fledgling

REAUME

THE GREAT HORNED OWL HAS AN impressive north-to-south range in the Americas. It can be found in an amazingly varied number of habitats from the sub-Arctic areas of North America all the way to southern parts of South America. Often described as the tiger of the sky, Great Horned Owls are powerful predators that take a wide variety of prey, from swans to skunks, herons to hares, Osprey to other owls and even domestic cats (yet another reason to keep your cat indoors). Major Bendire (1892, in Bent) "reports a nest that contained a mouse, a young muskrat, two eels, four bullhead catfish, a woodcock, four Ruffed Grouse, one rabbit and eleven rats. The food taken out of the nest weighed almost 18 pounds." Great Horned Owls in our area tend to focus on small rodents and Eastern Cottontails. As these prey species are often attracted to bird seed under feeders at night, many sleeping homeowners have Great Horned Owls hunting in their backyards.

SIZE

Length > Red-tailed Hawk
Wingspan < Red-tailed Hawk

Adult

LYNCH

LYNCH

Adult showing nictitating membrane, a protective eyelid.

ADULT

Ear tufts (can be flattened and hard to see, but are usually quite prominent) • *orange-brown facial disks with dark border* • yellow eyes • whitish eyebrows and mustache • *white throat patch* (can be concealed) • mottled brownish gray back • buffy underparts with *thin, dark barring* • short tail.

ADULT

Buffy underwings with a dark comma on wrist.

LISTEN FOR

The common call is a deep *hooo hoo-hoo hoooooo hoooooo.*

COMPARE TO

Long-eared Owl.

NATURE NOTES

For such a large bird, Great Horned Owls can be incredibly silent. I did an owl prowl early one winter morning and called for a Great Horned Owl in an area where a pair usually has a territory. I never got a response, but when I turned to leave, a Great Horned Owl flew away from the bare tree I was standing under. It had come in to my calls and landed above my head without my hearing it.

RANGE

Resident year round

Snowy Owl

Bubo scandiacus

Adult male

McCAW

First year female

McCAW

THE SNOWY OWL IS A WELCOMED winter visitor to the Great Lakes region. A very large owl, the Snowy is a favorite bird to watch for in the open fields of agricultural areas, but may be found in cityscapes as well. Most of the "snowies" that come this far south from their high Arctic breeding areas are heavily marked first years. Sometimes, though, an almost pure white adult male Snowy Owl may be seen perched on a fence post or telephone wire. Many of these owls stay in their Arctic homes all year round, but "irruptions" may occur in the south when major northern food sources such as lemmings become scarce. Snowy Owls are true nomads; they move to where they find prey. And they can really move. One Snowy fitted with a radio transmitter moved from a point in Siberia to a point in Arctic Canada, a distance of 1,800 miles (2,900 km), in 48 days. If a high abundance of prey can't be found during the breeding season, the owls may not even attempt to nest that year.

SIZE

Length > Red-tailed Hawk
Wingspan > Red-tailed Hawk

First year female

McCAW

ADULT MALE

All white body · most have some sparse black or dusky spots (individuals at least three to four years old may be pure white) · yellow eyes · short tail.

FEMALE & FIRST YEAR

Similar to adult male except for · *variable black barring on crown, back, wings, breast, belly and tail* (first year females have the most) · *facial disks still white.*

ADULT & FIRST YEAR

White underwings with variable barring.

LISTEN FOR

Usually only the males hoot, a booming *hoo* or short series of hoots that has been said to be heard up to 7 miles (11 km) away. Owls on their wintering grounds may grunt or whistle.

COMPARE TO

Barn Owl.

NATURE NOTES

The oldest known identifiable bird species in cave art is the Snowy Owl.

RANGE

■ Breeding only
□ Wintering only

Northern Hawk Owl

Surnia ulula

Adult

McCAW

Fledgling

McCAW

LIKE ITS NAME SUGGESTS, THIS boreal inhabitant combines owl characteristics with a hawklike body. Built for speed, the Northern Hawk Owl uses its short wings and long tail to do low, fast attacks on prey. It is also adept at listening from a perch and catching voles through a layer of snow. In winter, the Northern Hawk Owl sometimes comes south of its normal northern boreal forest home. Its mostly diurnal habits and remarkable tameness often allow birders to get very good looks.

SIZE

Length = American Crow
Wingspan = Rock Dove

Adult in flight

McCAW

🦅 ADULT

Light gray or white facial disks with *thick black borders that extend to the eyes* • yellow eyes • brownish gray crown with small white spots • *white and black lines on sides of the head* • brownish back with white mottling • *white underparts with rufous or brown barring* • *long tail*.

FLIGHT TRAITS

Fast flapping, like an accipiter, but can also hover like a kestrel.

LISTEN FOR

A whistled trill (similar to the display "call" of a snipe).

COMPARE TO

Cooper's Hawk, Sharp-shinned Hawk.

NATURE NOTES

Will do a distraction "broken wing" display to lead potential predators from fledged young.

Adult

McCAW

RANGE
▪ Resident year round

Barred Owl

Strix varia

Adult SMALL

Adult FLYNN

THE BARRED OWL'S DARK EYES set it apart from all of the other eastern North American owls except for the Barn Owl. Their wise look prompted Thoreau (1858) to write, "Solemnity is what they express, – fit representatives of the night." An inhabitant of often densely wooded areas, the Barred Owl is equally at home in southern swampland and northern boreal forest. Its loud, distinctive hooting has enhanced many campers' evenings. These hoots can sometime evolve into a maniacal series of notes that have given rise to a variety of other names for this species, such as Crazy Owl and Laughing Owl.

Adult

LYNCH

SIZE

Length > Red-tailed Hawk
Wingspan> American Crow

ADULT

Dark eyes • facial disks with brown border
• whitish or gray eyebrows and mustache •
brown upperparts with white mottling
• white or buffy underparts • *barred breast* •
thick dark streaks on belly • short tail.

ADULT

Buffy underwings with dark comma on wrist.

LISTEN FOR

A throaty series of deep hoots in the pattern
*Who cooks for you? Who cooks for
you-all?* Can also do a series of laughing
hoots and screams.

COMPARE TO

Great Gray Owl.

NATURE NOTES

This species will use specially made nest boxes.

RANGE

Resident year round

McCAW

Adult

McCAW

Adult

THE GREAT GRAY OWL'S LARGE SIZE, huge facial disks and penetrating stare give it an incredible presence. They can also be quite tame, which allows observers to watch their behavior. Though birders in the northern Great Lakes area may be able to see these owls in their resident boreal forest homes, most observers must wait for the winters when some Great Gray Owls come south. These owls are the biggest looking owls in North America, but they may weigh much less than a Snowy or Great Horned Owl. Because of this difference in bulk, Great Gray Owls tend to focus more on smaller prey than these other two species. Some populations may subsist almost entirely on voles.

Adult

McCAW

SIZE

Length < Bald Eagle
Wingspan > Red-tailed Hawk

ADULT

Large facial disks with dark border and thin, dark concentric circles • small yellow eyes • light gray eyebrows and mustache • broken white line across throat (can be hard to see) • body gray with dusky mottling.

ADULT

Gray underwings with no dark comma on wrist.

LISTEN FOR

A series of evenly spaced, low, booming hoots.

COMPARE TO

Barred Owl.

NATURE NOTES

Great Gray Owls have been known to plunge through 1½ feet (45 cm) of snow to catch rodents that they could hear from above.

RANGE

Resident year round

Adult — McCAW

Adult in defensive posture — FLYNN

LOOKING LIKE A MINIATURE replica of the Great Horned Owl, the Long-eared Owl can be distinguished by its smaller size, vertical black lines above and below its eyes and vertical streaks on its breast and belly. The Long-eared Owl makes quite a variety of calls. Thoreau described an encounter with a vocal owl of this species: "Just then I thought surely that I heard a puppy faintly barking at me four or five rods distant amid the bushes, having tracked me into the swamp, – what what, what what what. It was exactly such a noise as the barking of a very small dog or perhaps a fox. But it was the old owl, for I presently saw her making it."

Adult

McCAW

SIZE

Length < American Crow
Wingspan < American Crow

🦉 ADULT

Long ear tufts • yellow eyes • white eyebrow
• *orange-brown facial disks with dark border*
• *black vertical line through eye* • mottled
brownish gray back • buffy underparts with
dark barring and *streaking* and some white
mottling (female is darker overall) • short tail.

🪶 ADULT

Buffy underwings with a dark comma on the
wrist • similar to Short-eared Owl in flight.

LISTEN FOR

The male gives a series of evenly spaced, deep
hoots. This species has a wide array of other
calls that include squeals, hisses and high-
pitched barks.

COMPARE TO

Eastern Screech-Owl, Great Horned Owl,
Short-eared Owl.

RANGE
◼ Resident year round

Short-eared Owl

Asio flammeus

Adult

FLYNN

Adult

FLYNN

THIS STRIKING OWL WITH THE "TOO much mascara" look is often out hunting at dusk or even in broad daylight. This gives us a great opportunity to watch the owl's flight pattern and its interactions with other raptors and those of its own species. Short-eared Owls often roost and hunt together in the winter months. Seeing a group of 30 to 40 owls hunting over the same fields is an incredible sight. Their mothlike flapping is interspersed with long glides as they scan and listen for small mammals. Their open field or marsh habitat is often shared with Northern Harriers and the two species may dive at one another and steal each other's prey.

Adult

McCAW

SIZE

Length < American Crow

Wingspan = American Crow

ADULT

Whitish or buffy facial disks • yellow eyes • *black patch around eyes* • whitish eyebrows • very short ear tufts that are often hard to see • brown back and wings with buffy or whitish mottling • *whitish or buffy underparts with brown streaks* (females are usually darker overall) • short tail.

ADULT

Whitish or buffy underwings with a dark comma on the wrist • similar to Long-eared Owl in flight.

LISTEN FOR

A series of deep hoots. Also does high-pitched barking notes and a raspy *kee-yaaah*.

COMPARE TO

Long-eared Owl, Northern Harrier

NATURE NOTES

Short-eared Owls live in North America, South America and across northern Asia and Europe. They have also colonized some small islands. They have been recorded flying over the sea almost 700 miles (1,100 km) away from land.

RANGE

■ Breeding only
■ Resident year round
□ Wintering only

Boreal Owl

Aegolius funereus

Adult   McCAW

Fledgling   LYNCH

LIKE THE GREAT GRAY OWL AND Northern Hawk Owl, the well-named Boreal Owl sometimes moves south of its boreal forest habitat in times of winter prey shortages. However, unlike these other two owls, the Boreal Owl usually hunts at night and hides in dense conifers during daylight hours and so it is not often seen. When this owl becomes active at night to hunt, it uses hearing to find much of its prey. The Boreal Owl has one ear opening that is higher than the other. This allows the owl to hear sounds in both a horizontal and a vertical plane, which helps it pinpoint where prey sounds are coming from. This is a great advantage when hunting in the dark or finding prey hiding under a layer of snow. In varying degrees, many other owl species have this adaptation as well, but it is most pronounced in the Boreal Owl's skull shape.

Adult

LYNCH

SIZE

Length = American Robin
Wingspan > Mourning Dove

🦉 ADULT

White or grayish facial disk with thick, broken border that extends to the eyes on the top and the bottom • yellow eyes • white eyebrows • *dark forehead with white spots* • brown back with white spots • white underparts with thick, mottled streaks • short tail.

🦉 FLEDGLING

Dark brown head, facial disks, breast, wings and back • *white eyebrows* • whitish line on throat • white spots on wings • light brown belly and undertail coverts.

🦉 ADULT

White wing linings.

LISTEN FOR

A rapid series of short, hollow whistles.

COMPARE TO

Northern Saw-whet Owl.

NATURE NOTES

The size difference between females and males is more pronounced in this species than any other owl mentioned here. Small males may weigh less than 3 ounces (90 g) whereas large females may weigh more than 7 ounces (200 g). Boreal Owls often cache extra prey. In the winter, when this cached prey freezes, the owls must sit on the prey to thaw it out so it can be eaten.

RANGE
▮ Resident year round

Northern Saw-whet Owl

Aegolius acadicus

Adult

McCAW

Fledgling

LYNCH

YOU CAN'T REALLY LOOK AT A Northern Saw-whet Owl without thinking, "Wow, is that ever cute!" These are the smallest owls in eastern North America and their tameness is astounding. I found my first one by almost hitting it with my forehead while I was ducking under a conifer branch along a hiking trail. Instead of flying away, it first stretched itself vertically, then slowly lowered its body into a ball. While I was sketching it, it turned its head in the direction of a mobbing chickadee. I slowly held up my pencil and the bird followed it back into its original position, allowing me to finish my sketch. If only more wildlife art subjects were that cooperative!

SIZE

Length < American Robin
Wingspan = Mourning Dove

⬛ ADULT

Brownish streaked facial disks • yellow eyes •
white eyebrows • *brown forehead with short,
thin white streaks* • brown back and wings with
white spots • white underparts with thick,
rufous-brown streaks • *white undertail coverts*
• short tail.

⬛ FLEDGLING

Dark brown head, facial disks, back, breast
and wings • *white eyebrows that join together
on forehead, forming a triangular patch* • some
white or buffy mottling on wings • buffy-
orange belly and undertail coverts.

⬛ ADULT

Whitish or buffy wing linings.

LISTEN FOR

A long series of whistled toots. Also a raspy
kreee-ah kreee-ah kreee-ah series that
sounds similar to the sharpening of a saw
(thus, "saw-whet" owl).

COMPARE TO

Boreal Owl, Eastern Screech-Owl.

NATURE NOTES

Older Northern Saw-whet Owl nestlings have
been known to feed bits of food to their
younger siblings. Though tiny, these owls can
kill prey up to the size of a Rock Dove.

McCAW

Adult

RANGE

■ Resident year round
■ Wintering only

Vagrants

THERE ARE OVER 40 species of birds of prey that have been recorded in the United States and Canada. The following species are rare occurrences in most of eastern North America. All of these (except Harris' Hawk) can be found in Florida.

White-tailed Kite

Elanus leucurus

The White-tailed Kite can be distinguished from other kites by its black shoulder patch and pure white tail. In flight, it has dark primaries and a black wrist patch. This species is usually found in the west coast states as well as southern Texas and southern Florida. In the Great Lakes region, it has been recorded in Wisconsin, Indiana and New York.

Adult   SMALL   Adult   WHEELER

Snail Kite

Rostrhamus sociabilis

This kite is a rare resident in parts of Florida. Its long, thin beak is used to feed on large, aquatic snails. Unlike the other kites in this book, the Snail Kite has wide, rounded wings and a black tail with a wide, white base. The adult male is shown here.

Adult male   WHEELER   Adult male   WHEELER

Harris's Hawk

Parabuteo unicinctus

Adult — WHEELER

Adult — WHEELER

Harris's Hawk is a bird of the southwestern U.S., where it often teams up with others of its species to hunt cooperatively. It has bright rufous shoulders and wing linings that contrast with its dark primaries and secondaries and the black and white tail. It has been recorded in Ohio and Wisconsin.

Short-tailed Hawk

Buteo brachyurus

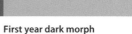

Adult light morph — WHEELER

First year dark morph — SMALL

The Short-tailed Hawk occurs in Florida and has two morphs. Unlike other hawks in this book, the dark morph is more common than the light morph. It is a shy species that hides in forested areas and is usually only seen while it is flying.

Crested Caracara

Caracara cheriway

Adult WHEELER

Adult McCAW

This relative of the falcons is usually found in Florida, Texas and Arizona, but has been recorded in Ontario and Minnesota. It is primarily a scavenger and can chase a single vulture off a carcass as well as steal food from buteos.

Burrowing Owl

Athene cunicularia

Adult McCAW

Adult McCAW

The Burrowing Owl is found over much of the western United States and southern prairie provinces, though its numbers are decreasing rapidly in some areas. It is also found in Florida. This species has been recorded in Ontario and all of the Great Lake states except Illinois and Pennsylvania.

Owl prowls & more

ONE OF MY FAVORITE birding activities is going on an owl prowl. Here are a few helpful tips on how to conduct your own.

WHERE?
- A wooded area bordering an old field is often a good place.
- During the day scout out good roosting areas by looking for owl pellets (regurgitated clumps of indigestible hair, bone and feathers) or white wash (owl poo on branches or tree trunks).

WHEN?
- Conduct prowls in the winter so owls with young are not disturbed.

HOW?
- You can use a tape recording of owl calls, but it is more fun trying to get an owl to respond from a sound that you make yourself.
- Try whistling through a bit of saliva on the back of your tongue for a really good Screech-Owl call.
- Making deep hoots in your throat will work for Great Horned and Barred Owls.
- Listen to a tape of owl calls and practice.

ETIQUETTE
- Once you get a good response, stop calling and listen (this makes it less stressful on the owl as it may think it has scared the "intruder" away from its territory).
- Stay still and watch for movement and silhouettes in the branches; some owls may stay silent but fly in closer to you.
- If possible, don't use a flashlight – this is a listening adventure!

After "talking" to an owl or watching the aerial prowess of a flying hawk, you may be wondering what you can do to help them. Well, the fact that you use this book has already helped; your interest alone can be very beneficial to all of nature. By learning more about hawks and owls, you have become even more connected to the natural world around you. When you share your interest with others, the growing support for wildlife and their varied habitats increases.

To be more active, you could volunteer at a raptor migration site counting the spring arrivals or fall departures. As well, you could volunteer your time to do owl surveys for bird monitoring programs or work at a local raptor rehabilitation center.

To help with nesting sites, you could build and monitor owl or kestrel nest boxes or help put up a nesting platform for Osprey. Information on how to construct and where to locate nesting boxes and platforms is available from a variety of sources including local naturalist groups, nature stores and the Wildlife Service.

McCAW

Raptor review
Put your skills to the test.

NOW IS YOUR CHANCE to test your hawk and owl identification skills. The following 24 photos are only numbered so that you can try to identify each species. Then you can go to the end of this section to check your answer and to read suggestions on how to identify this hawk or owl. Feel free to use the rest of the book when trying to identify each bird – after all, this is a field guide to be used in the field! Doing this review will not only get you more familiar with this book, it will be good practice on field marks, anatomy, shape and behavior. Good luck!

REAUME

REAUME

EGRESSY

4

REAUME

5

REAUME

6

BECK

7

REAUME

8

McCAW

9

BECK

10

REAUME

11

BECK

12

McCAW

13

EGRESSY

14

BECK

15

FAIRBAIRN

16

FAIRBAIRN

EGRESSY

EGRESSY

HOLDEN

20

21

22

23

HOLDEN

24

McCAW

1 The forward facing eyes and facial disks tell you that this is an owl, but which one is it? Young owls such as this one can be difficult to identify because they may leave the nest before they lose their downy look. In this case, the owl doesn't have its ear tufts yet either. This young **Great Horned Owl** does have the adult's black-bordered chestnut facial disks. While young Long-eared Owls would have this feature, too, the Great Horned has a more oblong face whereas the Long-eared would have a more circular face. These two species are quite different in size (the Great Horned is much bigger) but when threatened, as this owl is, size may be hard to judge. An owl may fluff itself up and spread its wings to look larger and more foreboding to a potential predator.

2 In contrast with the above Great Horned Owl, this owl is trying to make itself be less noticeable instead of trying to make itself look larger. **Eastern Screech-Owls** will often squeeze themselves into an upright position and point their ear tufts as high as they can. This makes them look like a broken branch so they can hide from predators during the day. The Eastern Screech-Owl is our only very small owl that has ear tufts. Grey morph Eastern Screech-Owls, such as this one, are more common in the northern and southern parts of their range. The red phase is more common in the central part of their range.

3 Just before this hawk took off, it left you with a lot of field marks to remember. You noticed a very dark wrist patch, a black belly band and a thick dark band at the end of its tail. This combination of features makes this a light morph **Rough-legged Hawk**. Review the identification features of this species (pages 68-71) to be ready to identify a dark morph Rough-legged. The dark morph isn't as common as the light morph, but it is a good idea to be prepared anyway!

4 The relatively large head of this small raptor suggests that this is a falcon. The setting coincides with the preferred habitat of the American Kestrel, which likes open field areas. While this bird does look like a female kestrel at first glance, there are a few key field marks that don't really fit. A female kestrel would have barring on her back, a greyish cap and a dark moustache and sideburns on her face. But more importantly, the tail is wrong for a kestrel. A female kestrel would have a finely barred tail, not a tail with thick dark bars separated with thin light bars. This feature helps us know that this is a **Merlin**. This individual is very light overall and might be the western subspecies of **Merlin**, even though this photo was taken in central Florida.

5 You can't see the face of this bird. Unfair, you say! But there will be many times when your subject will take off before you are able to see all of the features you'd like. This raptor has a long tail, some barring on its underparts and some mottling on its brown back. These field marks may suggest an accipiter (Sharp-shinned Hawk, Cooper's Hawk or Northern Goshawk), but the head is all wrong. It is too large and has dark and light markings, making this a **Northern Hawk-Owl**. Like some hawks, this mostly diurnal owl hunts on the wing and thus has an accipiter-like body shape. The markings on the back of its head are like those on its face. This adaptation makes it appear that the bird can see in any direction and might dissuade a potential predator or mobbing crow from getting too close.

6 Isn't it always the way? Just as you focus your binoculars on a perched hawk, it takes off. While it was hard to get a good look at the shape of this bird, you did notice a rather short tail and long, broad wings, leading you to think it was a buteo. And you saw some good field marks! The best was the two thick tail bars separated by a thick white bar. This rules out Red-shouldered Hawk, which would have several thick dark tail bars separated by thin white bars. This hawk also has a dark trailing edge to its otherwise mostly plain and pale underwings, making it a light morph **Broad-winged Hawk**. This is our smallest buteo though size in a quick encounter can be hard to judge.

7 Dark eagles are always difficult for beginners. I can remember constantly trying to figure out if what I was watching was an immature Bald Eagle or a Golden Eagle. Luckily, this bird is showing us the key feature to watch for. The wing linings of this young **Bald Eagle** have white mottling. Golden Eagles may have white at the base of their flight feathers, but their wing linings are dark. This youngster is "branching." This means it has left the nest but is not able to fly well yet. Here it is holding tightly to a branch and letting the wind move through its wings as it gets ready for its first big flight.

8 The hooked beak of this "raptor" can fool a beginner. This **Northern Shrike** is just as much of a hunter as our hawks and owls are, but it is a songbird and is more closely related to a vireo than a hawk. Its black mask and small size (the same length as an American Robin) are usually enough to set it apart from the others in this book. In winter, Northern Shrikes may become a backyard bird predator, terrorizing your bird feeder visitors just like Cooper's Hawks and Sharp-shinned Hawks do.

9 Hmmm, this is a tricky one! The raptor is dark all over and the first thing that should jump to mind is an eagle or some of the melanistic buteos that we have. The most common buteo would be a dark Rough-legged Hawk, and much less common dark Swainson's, Ferruginous, Red-tailed or Broad-winged Hawks. But here is where experience with overall shape comes into play. This bird just doesn't look "buteoish" or "eaglish." Shape takes a lot of practice, but at some point you will have a little voice in your head saying that this bird really doesn't look like a regular hawk. And it isn't. This dark morph **Gyrfalcon**'s large eye, round head and overall shape make it look "falconish." Hopefully this bird takes off and you will be able to see its pointed wings and fast and powerful flight. Don't worry if you found this image especially difficult, the more raptors you look at, the better you will be at identifying them. I promise!

10 While hiking at a park during your vacation to Florida, you come across this large bird. You can tell it is an owl, but you have a hard time seeing all of the bird to catch all of its field marks. Luckily there is a very important field mark staring you in the face – the owl's eyes. This owl has dark eyes and, thus, can only be one of two of our eastern owls. Because the owl isn't whitish or creamy overall and lacks a heart-shaped face, you rule out Barn Owl. That leaves the **Barred Owl** and you can just make out the streaks on its belly to confirm your identification.

11 Wow! Can you imagine seeing an owl just over your head like this? It certainly is an unforgettable experience. And this is the species that you would most likely see in this type of situation. Short-eared Owls often hunt in the late afternoon and sometimes in groups. I've seen 30–40 hunting or perched in the same open field. And when one eventually flies over your head and gives you that look it makes you happy that you aren't a meadow vole! This **Short-eared Owl** can be identified by its "too much mascara" eyes, breast streaks and overall buffy coloration. Its bouyant and butterfly-like flight pattern is important as well.

12 While enjoying the activity at your bird feeder, all the birds suddenly take off and disappear. Well, almost all the birds. The woodpeckers are still there, but they have all jumped to the same side of whatever tree they are on. This not only gives you a clue that a predator is around, but it tells you which direction to look. Woodpeckers will keep their tree trunk between them and the predator, so when you look the opposite direction away from the woodpeckers, you see this hawk perched nearby. The reddish barring on the underparts, the barred tail and the bright red eye tells you that this is either an adult Cooper's Hawk or an adult Sharp-shinned Hawk. The length of the feathers of the folded tail are all the same (or almost the same) length, so it is a **Sharp-shinned Hawk**. As well, the head of this bird is very dove-like and less "fierce-looking" than a Cooper's Hawk. These two species take a lot of practice but backyard feeders often give us a few good looks of each species every year.

13 This puffy ball of feathers is one of the most sought after owls in North America. The **Boreal Owl** is a northern species that sometimes comes south in the winter. It can be hard to tell apart from the more common, but still seldom seen, Northern Saw-whet Owl. The Boreal has a dark outer edge to its mostly white facial disks which includes a line that touches each eye. It also has white spots on its forehead as opposed to the white streaks of a Northern Saw-whet. Both of these species are rarely seen outside of dense conifers or shrubby thickets during the day.

14 So, what have we to work with here? Thick dark tail bars and reddish breast barring are the most prominent field marks. Missing field marks are important, too. There aren't any distinctive wrist patches or markings, no patagial patch and no belly band. The reddish breast barring alone narrows this to adults of Broad-winged Hawk, Red-shouldered Hawk, Cooper's Hawk and Sharp-shinned Hawk. The tail bars rule out the first two; they are neither two thick dark bars separated by a thick white bar (Broad-winged) or thick dark bars separated by thin white bars (Red-shouldered). There are many thick light and dark bars and the tail is too long to belong to a buteo like a Red-shouldered or Broad-winged. These two species would have wider, longer wings, too. The tail's fairly prominent white terminal bar suggests Cooper's, but this field mark isn't always conclusive on its own. The outer tail feathers do look shorter than the inner ones, another Cooper's field mark, but this is hard to see. But, the very straight leading edge to the wings and the large protruding head also help us decide that this bird is, indeed, a **Cooper's Hawk**. And you would know this if you looked on page 45 because it is the same photo (just checking to make sure you are using the book!).

15 While this hawk is quite a distant one, you can still see all you need to know to identify it. The wings are frequently held in a dihedral or "v"-shape and there is a prominent white rump. The orangish coloration on the breast and belly of this **Northern Harrier** means that it is a first year bird. This species is most commonly seen hunting low over fields and open wetlands.

16 Even when this bird's wings are fully stretched out, they still look pointed, helping us consider that this might be a falcon. When this speedy raptor comes in for a landing, you can see one of its best field marks. The dark underwings or "arm pits" are a distinctive marking of the **Prairie Falcon**. This bird is much more common out west but there are very rare reports of individuals showing up in the east during migration.

17 There aren't many field marks to go on for this one. The bird has a mottled belly band and plain breast. Its yellow eyes show it is a young bird. It is bulky overall and looks like it should be a buteo, which it is. Only Red-tailed Hawks and Rough-legged Hawks have distinct belly bands. The tail is hard to see, but it definitely doesn't have a thick dark subterminal band, so it can't be a Rough-legged Hawk. Yes, that leaves us with a first year **Red-tailed Hawk**. Remember to take close looks at all those Red-tails that you find. Even though you know they are Red-tails, getting familiar with field marks other than the red tail will pay off with young birds such as this one.

18 Here is another young bird with yellow eyes. This one isn't as bulky looking as the Red-tail above. The long banded tail isn't patterned correctly for any of our buteos and it doesn't have facial disks like a harrier. That leaves us with the accipiters: Northern Goshawk, Cooper's Hawk and Sharp-shinned Hawk. Because the bird has yellow eyes and is brown and white overall, we know that it is a first year bird. The pattern on the underparts is helpful for separating the 3 first year accipiters. A Northern Goshawk would have thicker streaks that would be found from the breast all the way to the far end of the undertail coverts. A Sharp-shinned Hawk also has thicker streaks that go farther down, though not right into the undertail covers like the goshawk. This first year **Cooper's Hawk** has thin streaks that stop in the belly area. As well, if you look closely at the tail, you can see that the outer tail feathers are much shorter than the central tail feathers, another great way to tell a Cooper's from a Sharp-shinned.

19 When hawks fly over snow on a bright day, you can often get great looks at their underparts because the snow reflects the light upwards. This is the reason this hawk is so wonderfully illuminated. And this gives us a great view of the needed characteristics to identify it. This bird has rusty barring on its breast and wing linings and a very dark trailing edge to the wings. There are two buteos with this combination, so we must now look at the tail. This hawk has very thick tail bands separated by thin white bands, making this an adult **Red-shouldered Hawk**. Most Red-shouldered Hawks leave the cold north during the winter, but sometimes a few stay and show off their brilliant colors above the snow.

20 While this is a bit of a grizzly shot, it shows the true nature of raptors: they are predators and must kill to survive. And what this raptor is carrying is a clue to its identity. The prey is a small duck called a Bufflehead and the predator was once known as the Duck Hawk. Its pointy wings, thick black sideburns and finely barred belly, underwings and tail reveal it to be a **Peregrine Falcon**. Peregrines are the sultans of speed and so they must do most of their hunting in the open, making swimming waterfowl a popular target.

21 Another hidden owl. This one does show its large ear tufts and chestnut facial disks, though, so you know that it is either a Great Horned Owl or a Long-eared Owl. These two can be hard to tell apart, especially if you can't get a good feel for how big the owl is. But there is something here that you can note. If you look closely, you can see that the breast feathers on this bird have horizontal bars as well as vertical streaks. Since Great Horned Owls only have horizontal barring on their breasts, this helps you conclude that you have found a **Long-eared Owl**.

22 If only raptors would freeze in the air like this so we could get clear looks at all of their field marks! But, since that won't happen, be sure to take into account how the bird is moving as it goes by. Is it fast or slow? Is it up high and circling, or low and diving, or somewhere in between? In this case, the bird was in hunting mode and trying to grab an Ivory Gull. This is a grey morph **Gyrfalcon**. Note the pointed wings (though not quite as pointed as a Peregrine's) and the moustache. It also has a finely barred tail and wings.

23 Unfair again! Yes, this is totally unfair, but before you cross me off your Christmas card list try to figure out what is going on in this photo. The perched bird appears to be an owl with ear tufts, but what is the flying bird? While crows might harass an owl during the day, they are way too smart to do so in the evening when it becomes the owl's advantage. This round-headed flier is actually another owl, a **Short-eared**, mobbing a **Great Horned Owl**. This may happen during the breeding season when the Great Horned gets too close to the Short-eared's nest. In this case, however, the Short-eared is trying to get the Great Horned to leave its winter hunting territory. Note that when owls fly they have such large wings that they can look much bigger than they really are. In this photo, this makes the Short-eared Owl appear only slightly smaller than the Great Horned when it really is significantly smaller.

24 Yes, this is a **Red-tailed Hawk**. That tail gives it away. But this shot is my final reminder that getting to know this hawk will be your best way of identifying other large raptors. Make sure you can accurately point out the patagial marks, commas, subterminal tail band and belly band here as well as knowing where the wing linings, undertail coverts and trailing edges of the wings are. Now get outside and practice!

References

BECAUSE OF THE THRILL of hawk-watching and the popularity of owls, there are many great references for birders to utilize. For hawks, *A Photographic Guide to North American Raptors* by Wheeler and Clark and *Hawks in Flight* by Dunne, Sibley and Sutton are musts for the serious raptor watcher. Wheeler is also soon to come out with another guide called *Raptors of Eastern North America*, which promises to be super, too. For owls, *North American Owls: Biology and Natural History* by Johnsgard has a lot of information for those who want more detail. As well, *Owls of the World: Their Lives, Behavior and Survival* by Duncan covers many topics and is a good read. *The Birds of North America*, edited by Poole and Gill, includes heaps of information on hawks and owls that has been collected by the scientific community.

"Aging Bald Eagles," by William S. Clark, *Birding*, February 2001.

Annotated Checklist of the Birds of Ontario, second edition by R.D. James, 1991, Royal Ontario Museum.

Atlas of the Breeding Birds of Ontario by M.D. Cadman, P.F.J. Eagles and F.M. Helleiner, 1987, University of Waterloo Press.

The Audubon Society Encyclopedia of North American Birds by John K. Terres, 1991, Wings Books.

The Birder's Handbook by P.R. Ehrlich, D.S. Dobkin and D. Wheye, 1988, Fireside/Simon and Schuster Inc.

The Birds of Canada – revised edition by W. Earl Godfrey, 1986, National Museum of Canada.

The Birds of North America series published by The Academy of Natural Sciences of Philadelphia and The American Ornithologists' Union.

Check-list of North American Birds, 1998, by The American Ornithologists' Union.

The Dictionary of American Bird Names, revised edition by E.A. Choate, 1985, Harvard Common Press.

Field Guide to the Birds of North America by the National Geographic Society, 1987.

Forest Raptors and their Nests In Central Ontario: A Guide to Stick Nests and their Uses, 1998, Queen's Printer for Ontario.

A Guide to Field Identification: Birds of North America by C.S. Robbins, B. Bruun and H.S. Zim, 1966, Golden Books.

Hawks in Flight by P. Dunne, D. Sibley and C. Sutton, 1988, Houghton Mifflin Company.

Hawks of Holiday Beach by A. Chartier and D. Stimac, 1993, Holiday Beach Migration Conservatory.

How to Spot Hawks and Eagles by C. Sutton and P.T. Sutton, 1996, Houghton Mifflin Company.

John James Audubon: Writings and Drawings, edited by C. Irmscher, 1999, The Library of America.

Life Histories of North American Birds of Prey, edited by A.C. Bent, 1937 and 1938, U.S. National Museum.

"The Magic of Snowy Owls" by Lynne Warren, National Geographic, December 2002.

National Audubon Society Field Guide to North American Birds – eastern region by J. Bull and J. Farrand, Jr., 1994, Alfred A. Knopf, Inc.

North American Owls: Biology and Natural History by P.A. Johnsgard, 1988, Smithsonian Institute Press.

Owls by Day and Night by H.A. Tyler and D. Philips, 1978, Naturegraph Publishers Inc.

Owls in Folklore and Natural History by V. Holmgren, 1988, Capra Press.

Owls of the World: Their Lives, Behavior and Survival by J. R. Duncan, 2003, Firefly Books.

Owls of the World by R. Hume, 1991, Running Press.

Owls. Their Natural & Unnatural History by J. Sparks, 1970, Taplinger Publishing Company.

Peterson Field Guides: Advanced Birding by Kenn Kaufman, 1990, Houghton Mifflin Company.

Peterson Field Guides: Eastern Birds by R.T. Peterson, Houghton Mifflin.

Peterson Field Guides: Hawks by W.S. Clark and B.K. Wheeler, 1987, Houghton Mifflin.

Peterson Natural History Companions: Lives of North American Birds by Kenn Kaufman, 1996, Houghton Mifflin.

A Photographic Guide to North American Raptors by B.K. Wheeler and W.S. Clark, 1995, Academic Press.

Raptors of Eastern North America by B.K. Wheeler, 2003, Princeton University Press.

Raptors: North American Birds of Prey by N. and H. Snyder, 1991, Raincoast Books.

Raptors of the World by J. Ferguson-Lees and D.A. Christie, 2001, Houghton Mifflin Company.

Seasonal Status of Birds: Point Pelee National Park and Vicinity compiled by J.R. Graham, 1996.

The Sibley Guide to Birds by D.A. Sibley, 2000, Alfred A. Knopf.

Stokes Behaviour Guide: Bird Behaviour Vol. 3 by Donald and Lillian Stokes, 1989, Little, Brown & Company.

Stokes Field Guide to Birds: Eastern Region by Donald and Lillian Stokes, 1996, Little, Brown & Company.

Studer's Popular Ornithology: The Birds of North America edited by J.H. Studer, 1881, Harrison House.

Thoreau on Birds by Henry David Thoreau, 1910, Beacon Press.

Owl comparison chart

Great Horned Owl · page 90   McCAW

Long-eared Owl · page 100   McCAW

Eastern Screech-Owl · page 88   SMALL

Short-eared Owl · page 102   FLYNN

Northern Saw-whet Owl · page 106   McCAW

Boreal Owl · page 104   McCAW

Barred Owl • page 96   SMALL

Barn Owl • page 86   SMALL

Great Gray Owl • page 98   McCAW

First year Snowy Owl • page 92   McCAW

Northern Hawk Owl • page 94   McCAW

Adult male Snowy Owl • page 92   McCAW

WHEELER

First year Sharp-shinned Hawk

WHEELER

First year Cooper's Hawk

WHEELER

First year Northern Goshawk

WHEELER

Adult Sharp-shinned Hawk • page 40

FLYNN

Adult Cooper's Hawk • page 44

FLYNN

Adult Northern Goshawk • page 48

	Head	Breast & belly	Undertail coverts	Tail shape	Tail pattern	Legs
First year Sharp-shinned	Small	White with thick brown streaks	Plain white	Long & thin with a squarish tip, outside tail feathers similar in length to the others	Thick dark bands with thin white terminal band	Long & very thin
First year Cooper's	Large	White with thin brown streaks	Plain white	Long & thin with a rounded tip, outside tail feathers shorter than the others	Thick dark bands with fairly thick white terminal band	Long & stronger looking
First year Northern Goshawk	Size in proportion to body, whitish eyebrow	Buffy or whitish with thick brown streaks	White with dark streaks	Long & broad	Thick, dark jagged bands, each with a thin white border on upper surface	Strong-looking
Adult Sharp-shinned	Small	White with thick rusty-orange barring	Plain white	Long & thin with a squarish tip, outside tail feathers similar in length to the others	Thick dark bands with thin white terminal band	Long & very thin
Adult Cooper's	Large	White with thick rusty orange barring	Plain white	Long & thin with a rounded tip, outside tail feathers shorter than the others	Thick dark bands with fairly thick white terminal band	Long & stronger looking
Adult Northern Goshawk	Size in proportion to body, whitish eyebrow	White with very thin gray barring	Plain white	Long & broad	Thick dusky bands	Strong-looking

Note: more than one characteristic is often needed to correctly identify accipiters.

First year Northern Goshawk — WHEELER

Adult Northern Goshawk • page 48 — WHEELER

First year Cooper's Hawk — WHEELER

Adult Cooper's Hawk • page 44 — BECK

First year Sharp-shinned Hawk — WHEELER

Adult Sharp-shinned Hawk • page 40 — WHEELER

	Head	Breast & belly	Undertail coverts	Wing shape in a soar	Flapping style	Tail shape	Tail pattern
First year Sharp-shinned	Small	White with thick brown streaks	Plain white	Short & rounded, often pushed forward	Fast & floppy	Long & thin with a squarish tip, outside tail feathers similar in length to the others	Thick dark bands with thin white terminal band
First year Cooper's	Large, often projecting well beyond leading edge of wings	White with thin brown streaks	Plain white, often extends into lower belly	Straight, longer than others	Stiff	Long and thin with a rounded tip, outside tail feathers shorter than others	Thick dark bands with fairly thick white terminal band
First year Northern Goshawk	Size in proportion to body, whitish eyebrow	Buffy or whitish with thick brown streaks	White with dark streaks	Broad & often more pointed	Deep & buteolike	Long & broad	Thick, dark jagged bands, each with a thin white border
Adult Sharp-shinned	Small	White with thick rusty-orange barring	Plain white	Short & rounded, often pushed forward	Fast & floppy	Long & thin with a squarish tip, outside tail feathers similar in length to the others	Thick dark bands with thin white terminal band
Adult Cooper's	Large, often projecting well beyond leading edge of wings	White with thick rusty-orange barring	Plain white	Straight, longer than others	Stiff	Long and thin with a rounded tip, outside tail feathers shorter than others	Thick dark bands with fairly thick white terminal band
Adult Northern Goshawk	Size in proportion to body, white eyebrow	White with very thin gray barring	Plain white	Broad & often more pointed than others	Deep & buteolike	Long & broad	Thick dusky bands

Note: more than one characteristic is often needed to correctly identify accipiters.

Broad-winged/Red-shouldered Hawk comparison chart – in flight

First year Broad-winged WHEELER

First year Red-shouldered WHEELER

Adult Broad-winged • page 54 HOLDEN

Adult Red-shouldered • page 52 HOLDEN

	Primary "window"	Primaries & secondaries	Wing linings	Tail shape	Tail pattern
First year Broad-winged	No	White with thick dusky band on trailing edge, some faint thin barring	White with some streaks	Very short	Fairly thick dark terminal band & many thin bands
First year Red-shouldered	Yes, but depends on lighting conditions	Faint black & white bars	White with some streaks	Longer	Many dark bands & thin white bands
Adult Broad-winged	No	White with thick black band on trailing edge, some faint thin barring	White with maybe some rusty barring	Very short	Two thick black bands separated by one thick white band
Adult Red-shouldered	Yes, but depends on lighting conditions	Black & white bars	White with rusty-orange barring	Longer	Black with two to three thin white bands

Hawk comparison chart – perched

WHEELER

Broad-winged Hawk • page 54

WHEELER

Sharp-shinned Hawk • page 40

McCAW

Red-shouldered Hawk • page 52

FLYNN

Cooper's Hawk • page 44

McCAW

Red-tailed Hawk • page 62

FLYNN

Northern Goshawk • page 48

WHEELER

Rough-legged Hawk • page 68

WHEELER

First year Northern Harrier • page 36

SMALL
Female American Kestrel • page 74

WHEELER
Merlin • page 78

SMALL
Peregrine Falcon • page 82

WHEELER
Osprey • page 24

WHEELER
First year Bald Eagle • page 32

McCAW
Turkey Vulture • page 20

WHEELER
Golden Eagle • page 72

SMALL
Bald Eagle • page 32

Hawk comparison chart – in flight

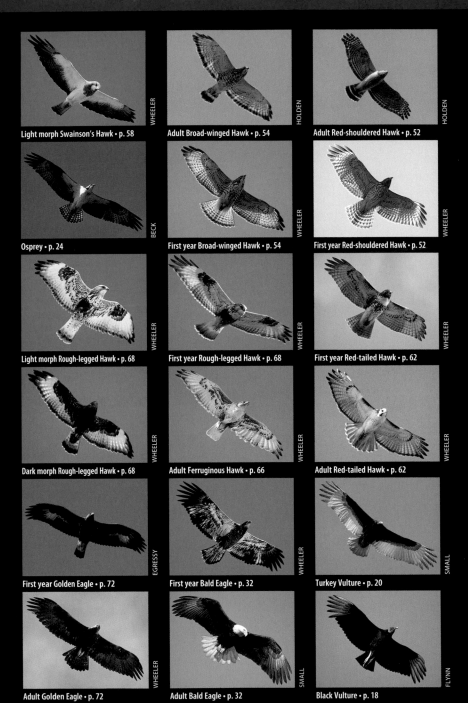

Light morph Swainson's Hawk • p. 58 — WHEELER

Adult Broad-winged Hawk • p. 54 — HOLDEN

Adult Red-shouldered Hawk • p. 52 — HOLDEN

Osprey • p. 24 — BECK

First year Broad-winged Hawk • p. 54 — WHEELER

First year Red-shouldered Hawk • p. 52 — WHEELER

Light morph Rough-legged Hawk • p. 68 — WHEELER

First year Rough-legged Hawk • p. 68 — WHEELER

First year Red-tailed Hawk • p. 62 — WHEELER

Dark morph Rough-legged Hawk • p. 68 — WHEELER

Adult Ferruginous Hawk • p. 66 — WHEELER

Adult Red-tailed Hawk • p. 62 — WHEELER

First year Golden Eagle • p. 72 — EGRESSY

First year Bald Eagle • p. 32 — WHEELER

Turkey Vulture • p. 20 — SMALL

Adult Golden Eagle • p. 72 — WHEELER

Adult Bald Eagle • p. 32 — SMALL

Black Vulture • p. 18 — FLYNN

Female Northern Harrier • p. 36 — WHEELER

First year Northern Harrier • p. 36 — WHEELER

First year Northern Goshawk • p. 48 — WHEELER

First year Cooper's Hawk • p. 44 — HOLDEN

First year Sharp-shinned Hawk • p. 40 — WHEELER

Adult Northern Goshawk • p. 48 — WHEELER

Adult Cooper's Hawk • p. 44 — BECK

Adult Sharp-shinned Hawk • p. 40 — WHEELER

Male Northern Harrier • p. 36 — WHEELER

Adult Mississippi Kite • p. 30 — WHEELER

First year Mississippi Kite • p. 30 — WHEELER

Gyrfalcon • p. 80 — BECK

Merlin • p. 78 — SMALL

Female American Kestrel • p. 74 — WHEELER

Prairie Falcon • p. 84 — WHEELER

Peregrine Falcon • p. 82 — WHEELER

Male American Kestrel • p. 74 — WHEELER

Index